T0191881

Monographs in Theoretical Computer Science. An EATCS Series

Series Editors

Books published in this series present original results or consolidated material of interest to the research community and graduate students. Manuscripts to be considered for publication should be coherent monographs rather than collections of articles. The series also contains high-level presentations of special topics.

Vittorio Bilò • Cosimo Vinci

Coping with Selfishness in Congestion Games

Analysis and Design via LP Duality

Springer

Vittorio Bilò
Dept. Mathematics and Physics
University of Salento
Lecce, Italy

Cosimo Vinci
Dept. Mathematics and Physics
University of Salento
Lecce, Italy

ISSN 1431-2654 ISSN 2193-2069 (electronic)
Monographs in Theoretical Computer Science. An EATCS Series
ISBN 978-3-031-30263-3 ISBN 978-3-031-30261-9 (eBook)
https://doi.org/10.1007/978-3-031-30261-9

This Springer imprint is published by the registered company Springer Nature Switzerland AG
The registered company address is: Gewerbestrasse 11, 6330 Cham, Switzerland

"Un bel problema, anche se non lo risolvi, ti fa compagnia se ci pensi ogni tanto."

Ennio De Giorgi

"Computer science is no more about computers than astronomy is about telescopes."

Edsger W. Dijkstra

"The mathematician's patterns, like the painter's or the poet's must be beautiful; the ideas, like the colours or the words, must fit together in a harmonious way. Beauty is the first test: there is no permanent place in the world for ugly mathematics."

Godfrey Harold Hardy

Preface

The fundamental aspect of the Theory of Computation is to understand the power and limitations of algorithms and, by consequence, of computers.

From a historical point of view, the first limitation considered in the development and analysis of algorithms has been the lack of unlimited computational power. This has lead to the birth of Complexity Theory, classifying problems with respect to the hardness of computing or, subordinately, approximating their solutions (Theory of Approximation Algorithms). The second limitation come under scrutiny has been the lack of knowledge of the input instance, which has given rise to the Theory of Online Algorithms. Finally, the third and last limitation to be considered has been the lack of coordination among the parties implementing the solution of a computational problem in a multi-agent system. Together with other questions, this has lead to the birth of Algorithmic Game Theory and, in particular, to the systematic study of the *Price of Anarchy* of non-cooperative games.

Some of the most basic and general solutions developed in the field of approximation and online algorithms, although focusing on different aspects of computation, share a unifying common feature: they can be naturally and effectively attacked by using instruments coming from the Theory of Linear Programming and Duality. *Primal-dual formulations*, in particular, have proved to hold the key of this power and one can refer to Vazirani [173] and Buckbinder and Naor [47] for applications to approximation and online algorithms, respectively.

Given this huge flexibility, it becomes natural to ask whether primal-dual formulations may also be used to successfully characterize the Price of Anarchy of games. It turns out that the answer is yes, as demonstrated by the seminal work of Vittorio Bilò [22], where previously known results for a fundamental class of non-cooperative games, called *congestion games*, were re-obtained by analysing suitable pairs of primal-dual formulations. Congestion games are perhaps the most studied and representative class of non-cooperative games, and are often elicited as the standard arena where to test and apply new solution concepts, modeling ideas and proof techniques.

Starting from this evidence, we have produced a series of works applying primal-dual formulations to answer a variety of questions in the realm of congestion games, often solving open problems, or widely generalizing previously known results. A big portion of these works was obtained during Cosimo Vinci's PhD program, and the resulting thesis was named the Best Italian PhD Thesis in Theoretical Computer Science of 2019 by the Italian Chapter of the European Association for Theoretical Computer Science. As part of the award, an agreement between the Italian Chapter and Springer established the publication of the thesis. This agreement has been subsequently modified and extended to produce the present monograph.

Although we left out some of the most involved and tedious constructions, the book contents are rather technical and require strong bases in mathematics and theoretical computer science. Thus, we believe that it is unlikely to be used as a textbook in undergraduate courses, but rather it may hopefully serve as a reference guide to postgraduate students and researchers who may be interested in deepen their knowledge on the fascinating field of the price of anarchy in congestion games and related topics.

We would like to conclude by pointing out that the techniques outlined in this book have been recently exported outside the realm of congestion games, see [40]. This should boost additional motivations to explore and master the power of primal-dual formulations as an investigation tool for characterizing the impact selfish behavior in non-cooperative games.

Acknowledgements

The content of this book is mainly based on the results achieved by the authors when Cosimo Vinci was a PhD student at the Gran Sasso Science Institute (GSSI), under the supervision of Vittorio Bilò. The authors would like to highlight that the writing of this book is not only the result of their research activity, but it has been possible thanks to the work and support of other people, whom the authors deeply thank.

First of all, the authors would like to thank Wayne Wheeler, Ronan Nugent, Priyanshi Peelwan, Naomi Portnoy and the editorial board of the EATCS Monographs book series, for their precious editorial work and the time they dedicated during all stages of writing the book. Furthermore, the authors wish to express individual thanks to all those who have contributed to the research work on which this book is based, with both their scientific and human support (see below).

VB: I have been studying the consequences of selfish behavior since 2003, when my then-PhD supervisor Michele Flammini told me about this research topic after a visit at IMT of Lucca. Since then, it has been a twenty-year pleasant trip during which I had the privilege to collaborate with some colleagues with whom I also developed long-standing friendship relations: Michele, Angelo Fanelli, Gianpiero Monaco, Luca Moscardelli, Ioannis Caragiannis, Marios Mavronicolas.

At a certain point during the trip, I was lucky to meet Cosimo and become his advisor. Although I could perceive his talent for research, what Cosimo achieved during his PhD went beyond my wildest dreams and the credit he got for his work is surely well deserved.

I believe that good-quality research is often the result of a good-quality state-of-mind. Thus, a special thank goes to my family: my parents, my brothers, my sisters-in-law, my nephews and to Valentina.

CV: First of all, I wish to express all my gratitude to my doctoral supervisor, Prof. Vittorio Bilò (from the University of Salento), with whom I co-authored several research works, including this book. Vittorio is a very brilliant scientist and professor and, at the same time, a very good person and friend. He has encouraged me to pursue a scientific career, and he has introduced me into the world of research, giving me a lot of inspiration. The precious work of Vittorio, his guidance, support and advices have been fundamental for my research activity, and I am deeply indebted with him.

I would like to thank Prof. Michele Flammini and Prof. Gianlorenzo D'Angelo (both from GSSI), with whom I have been collaborating for several years. In particular, I am grateful to them for their constant presence and guidance during my PhD studies (and all the subsequent periods), for the many fruitful discussions I had with them, for their kindness and willingness to help me for everything I needed.

I would like to thank Prof. Rocco De Nicola (from IMT Lucca), who has been the first coordinator of the PhD program in Computer Science at GSSI, Prof. Luca Aceto (from Reykjavik University), Prof. Pierluigi Crescenzi (from GSSI), Prof. Mattia D'Emidio (from University of L'Aquila), Prof. Ivano Malavolta (from Vrije Universiteit Amsterdam), Dr. Omar Inverso (from GSSI), Prof. Catia Trubiani (from GSSI), and the other professors and researchers met at GSSI, for the importance of their work and the support devoted to all PhD students.

I would like to thank Prof. Ioannis Caragiannis (from Aarhus University), whose brilliant results and research directions have inspired a substantial part of this book. He gave me the opportunity to work with him in Greece (when he was professor at Patras University), where I met other very good people, and to spend a wonderful period there, during which he has always been extremely kind and hospitable.

I would like to thank the other co-authors of some works which this book is based on: Prof. Diodato Ferraioli (from the University of Salerno), Prof. Vasilis Gkatzelis (from Drexel University), Prof. Gianpiero Monaco (from the University of L'Aquila), Prof. Luca Moscardelli (from the University of Chieti-Pescara).

I would like to thank Prof. Dimitris Fotakis (from National Technical University of Athens) and Prof. Marc Uetz (from the University of Twente), for carefully reviewing my doctoral dissertation (which this book is based on), and for their insightful comments. Furthermore, I would like to thank the committee members for my doctoral dissertation (Prof. Luca Aceto, Prof. Rocco De Nicola, Prof. Michele Flammini, Prof. Gianpiero Monaco and Prof. Giuseppe Persiano) for devoting their time to my dissertation defence.

I would like to thank the research groups in Computer Science I joined at the University of Salento (Prof. Vittorio Bilò and Dr. Antonio Caruso) and the University of Salerno (Prof. Vincenzo Auletta and Prof. Diodato Ferraioli).

I would like to thank my dear colleagues and friends met during my periods of study and/or research at GSSI, University of Salento, University of Salerno and University of L'Aquila.

In particular, regarding GSSI, I would like to thank (in alphabetical order by first name) Ahmed Abdelsalam, Alena Myshko, Alessandro Aloisio, Alessia Mastrangioli, Alessia Vannicelli, Alkida Balliu, Andrea Mandarano, Andrea Mazzon, Bojana Kodric, Carlo Caporali, Claudio Savarese, Clemens Grabmayer, Debashmita Poddar, Dennis Olivetti, Diego Vescovi, Eirini Lefaki Glynou, Elena Di Iorio, Eliana Di Giovanni, Emanuele Belotti, Emilio Incerto, Emily Catena, Fabrizia Di Stefano, Feliciano Colella, Francesca Alemanno, Francesca Ghinami, Francesco Cellinese, Gabriela Gorova, Gennaro Ciampa, Gian Luca Scoccia, Giovanna Varricchio, Grazia Di Giovanni, Hugo Gilbert, Lars Hientzsch, Lorenzo Severini, Luca Alasio,

Luca Giurina, Luigi Forcella, Marco Celoria, Manuel Mauro, Mariateresa Rossi, Martina De Sanctis, Massimo Bertolin, Matteo Tonelli, Mattia Manucci, Michele Aleandri, Michele Dolce, Mohamed Benyahia, Mora Durocher, Paolo Di Francesco, Piergiorgio Pilo, Raffaele Scandone, Raffaello Carosi, Roberto Boccagna, Ruben Becker, Simone Fioravanti, Stefano Ponziani, Stefano Ruberto, Tan Duong, Valentina Meschini, Valeria Raimondi, Yllka Velaj and Yuriy Zacchia Lun.

Regarding the University of Salento, I would like to thank (in alphabetical order by first name) Alberto Cioni, Alessandro Carbotti, Alessandro Melissano, Anna Romano, Antonio Costantini, Antonio Vitale, Carmela Peluso, Chiara Errico, Cristina Mancini, Edoardo Cleopazzo, Efrem De Pascalis, Emilio Falcicchia, Enrico Spada, Francesco Bisanti, Francesco Delle Donne, Francesco Lombardo, Gianfranco Colucci, Gianluca Matera, Lucaleonardo Bove, Leo Sergio, Luigi Gabrieli, Luigi Negro, Marco Castelli, Marco Guido, Massimiliano Gervasi, Massimo Frittelli, Pierluigi Bianco, Simone Cito, Veronica La Stella and Vincenzo Cambò.

Regarding the University of Salerno, I would like to thank (in alphabetical order by first name) Antonio Coppola and Leonardo Rundo.

Regarding the University of L'Aquila, I would like to thank (in alphabetical order by first name) Antonio Esposito, Giada Cianfarani, Graziano Stivaletta, Jon May, Simona D'Evangelista, Teresa Scarinci and Valeria Iorio.

I hope I haven't forgotten anyone! With them I have shared courses to attend, exams to pass, offices, but above all I have experienced good times and wonderful adventures (and I continue to do so).

Last but not least, I am grateful to my beautiful family, the people I love most (several of them have been mentioned above), and the loved ones who are no longer there. I dedicate this book to all of them, as their constant thought and closeness has always supported and accompanied me, everyday.

Contents

Part I
Introduction

Chapter 1
Coping with Selfishness in Congestion Games

"To feel much for others and little for ourselves; to restrain our selfishness and exercise our benevolent affections, constitute the perfection of human nature."

Adam Smith

1.1 The Impact of Selfish Behavior in Congestion Games

Since the end of the twentieth century, the computer science community has been interested in the study of complex systems populated by (numerous) rational agents interacting with each other, and in how their selfish behavior impacts on the overall social welfare. The study of these systems is related to the theory of *non-cooperative games* (Nash [133], Osborne [136], von Neumann and Morgenstern [177]), where selfish players choose a strategy from a set of alternatives with the aim of optimizing their own *utility function*, which depends also on the strategic choices of the others. In this setting, the notion of *pure Nash equilibrium* (Nash [133]), that is, a strategy profile in which no player gets a benefit when unilaterally changing her strategy, is widely adopted as the ideal solution concept, and several classes of games have been shown to posses such equilibria. Further generalizations of this concept, such as mixed Nash equilibria, correlated equilibria, coarse correlated equilibria and Bayes-Nash equilibria, have also been largely used and investigated (Aumann [10], Moulin and Vial [131], Nash [133]).

It has been known since the early fifties, with the famous prisoner's dilemma game, that an equilibrium is usually achieved at the expenses of the social welfare. In this sense, the computer science community has focused in quantitative aspects of non-cooperative games, by evaluating the quality of outcomes arising from selfish behavior. This interest, together with other new research directions in computational aspects of game theory, has lead to new challenges in this scientific area, and, in particular, to the birth of *Algorithmic Game Theory* (Nisan et al. [134], Papadimitriou [143]), which collocates at the intersection between computer science, mathematics and economics.

V. Bilò, C. Vinci, *Coping with Selfishness in Congestion Games*, Monographs in Theoretical Computer Science. An EATCS Series, https://doi.org/10.1007/978-3-031-30261-9_1

Efficiency Metrics in Non-Cooperative Games

To capture and measure the degradation of social welfare due to selfish behavior, Koutsoupias and Papadimitriou [113] proposed the *coordination ratio*, later popularized under the name of *price of anarchy* (Papadimitriou [143]), which is defined as the worst-case ratio between the social welfare of a pure Nash equilibrium and that of an optimal profile which could be enforced if players cooperated (*social optimum*). Other metrics are: (*i*) the *price of stability* (Anshelevich et al. [8], Schulz and Stier-Moses [164]), that is defined as the price of anarchy, but considers the best-case ratio, and (*ii*) the *competitive ratio of one-round walks starting from the empty state*, defined as the worst-case ratio between the social welfare of a profile obtained after a process in which players enter the game according to some arbitrary ordering and greedily choose an irrevocable strategy based on the choices of their predecessors only and the social optimum.

Congestion Games

The previous metrics have been estimated for several games. Among these, *congestion games* occupy a preeminent role. In a congestion game (Rosenthal [154]), there are a finite set of resources E and a finite set of players, and each player selects a set of resources from a collection of subsets of E. The utility that a player gets in a given strategy profile is equal to the sum of the costs of the resources that she selects, where the cost of each resource is a function (usually called *latency function*) that only depends on the number of players using the resource (the resource *congestion*). Many applications in real-life systems, such as routing on transportation or computer networks, machine scheduling, resource sharing, group formation and so on, are suitably modeled by congestion games. As examples, one may think to web users sharing limited resources (servers, connections, processors and so on), or drivers who want to move as fast as possible from a location to another along a road network. Some specific classes of congestion games have received a particular attention, because of their wide applicability to several problems coming from computer science and operations research. Among these, the subclass of load balancing games, in which each player can choose at most one resource, has been largely investigated. Indeed, load balancing games define a game-theoretic framework of machine scheduling problems, in which jobs/tasks (modeled by players) must be assigned to some machines/servers (modeled by resources).

1.2 How Can We Reduce the Impact of Selfishness?

The efficiency metrics defined above have been studied intensively in several variants of congestion games and they have been often shown to be significantly high.

Anyway, in several practical situations, selfish behavior can be partially controlled by an external authority having the goal of leading the game towards outcomes with reasonably good social welfare.

Example 1.1. As an example, consider the following problem related to traffic theory, which can be modeled as a congestion game. Suppose that there is a set of n drivers that, starting from a source s, want to reach a destination t in a road network; each driver wants to minimize the total travel time, and the travel time of each road is a non-decreasing function of the number of drivers running trough it, see Figure 1.1. It is easy to see that, for any $d \geq 1$ and $\varepsilon > 0$, there is a unique pure Nash equilibrium in this game, in which all drivers choose the upper road, for a personal cost of n^d. If we measure the social welfare by the sum of the travel time of all players, we have that the social welfare at equilibrium is equal to n^{d+1}, while simple arguments of calculus yield that the social optimum is achieved when only $\frac{n}{\sqrt[d]{d+1}}$ drivers take the upper road, for a social welfare of $n^{d+1}\frac{(d+1)\sqrt[d]{d+1}-d}{(d+1)\sqrt[d]{d+1}}+o(n\varepsilon)$. It follows that, by letting ε going to zero, the price of anarchy (indeed even the price of stability) of this game becomes $\frac{(d+1)\sqrt[d]{d+1}}{(d+1)\sqrt[d]{d+1}-d}$, which equals $4/3$ for $d = 1$ and grows arbitrarily large as d increases. For later reference, observe that, for $d = 1$, the social optimum is attained when $\frac{n}{\sqrt[d]{d+1}} = n/2$ drivers take the upper road.

This game was already considered by Pigou [150], and his studies anticipated various concepts of modern game theory. Indeed, it is a representative example of how selfish players may harm each other.

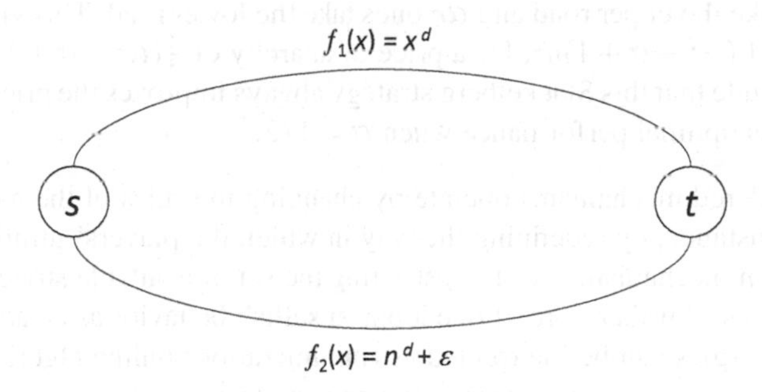

Fig. 1.1 The Pigou's network considered in Example 1.1.

In the following, we consider a couple of approaches that one can adopt to guide players toward better outcomes.

Taxation Mechanisms: A simple way to reduce the price of anarchy in this context is to impose a tax to each driver that runs through a road, thus defining a *taxation mechanism* (Pigou [150]). This approach has been also extended to

the whole class of congestion games and has been also generalized to other measures of social welfare and to different behavioral attitudes (for instance, one may or may not include taxes in the social welfare (*non-refundable* vs. *refundable taxes*) and/or consider a different sensitivity of players to taxes). For general congestion games, a tax can be defined as a function of the game's structure, and each player, when selecting a resource, is charged a tax that causes an increase in her perceived latency, thus guiding all players to reach better outcomes in terms of social welfare. Furthermore, taxes are applied locally by each resource, thus defining a decentralized and easily implementable mechanism, in which each resource is an autonomous computational entity.
With respect to the game defined in Example 1.1, consider the case of $d = 1$ and a taxation scheme imposing a toll of $n/2$ on the upper road. By virtue of this, function f_1 becomes $f_1(x) = x + n/2$. Thus, using this road remains profitable as long as its congestion does not exceed $n/2$. Hence, in the unique pure Nash equilibrium of the game, exactly half of the drivers take the upper road, thus implementing the social optimum. So, we can conclude that there exists a taxation mechanism inducing a price of anarchy of 1.[1]

Stackelberg Strategies: Another approach assumes that a central authority is able to control the strategies of a fraction $\alpha \in (0,1)$ of the players. The players to be controlled are chosen by an algorithm called *Stackelberg strategy* (Roughgarden [156]). The application of different Stackelberg strategies has been largely studied, and the resulting price of anarchy has been estimated as a function of α.
Let us consider again the game defined in Example 1.1, with $d = 1$. For any $\alpha \in (0,1)$, our Stackelberg strategy assigns all controlled αn drivers to the lower road. Choosing the upper road remains profitable for the remaining $(1-\alpha)n$ drivers, thus there is a unique pure Nash equilibrium in which exactly $(1-\alpha)n$ drivers take the upper road and αn ones take the lower road. This yields a social welfare of $(\alpha^2 - \alpha + 1)n^2$, for a price of anarchy of $\frac{4}{3}(\alpha^2 - \alpha + 1) \in (1, 4/3)$. We conclude that this Stackelberg strategy always improves the price of anarchy and yields optimal performance when $\alpha = 1/2$.[2]

All the considered mechanisms operate by changing the rules of the game in some manner, for instance, by redefining the way in which the players' utilities are computed (taxation mechanisms), or by restricting the set of available strategies (Stackelberg strategies). Furthermore, if one looks at selfish behavior as an adversarial input, these strategies can be interpreted as (minimization) online algorithms, whose competitiveness corresponds to some efficiency metric.

---

[1] The cost produced by taxation is not considered in the definition of the social welfare. This approach is referred in the literature as *refundable taxes*.

[2] Indeed, there is a better strategy, obtained by assigning the lower road exactly $\min\{\alpha n, n/2\}$ drivers, which yields an improved and optimal price of anarchy for any $\alpha > 1/2$. We choose not to illustrate this strategy to keep the description simple at these early introductory stages.

1.3 Scope of the Book

Congestion games constitute perhaps the most significant class of non-cooperative games, because of their effectiveness in modeling several real scenarios. Since the advent of Algorithmic Game Theory, characterizing the inefficiency of selfish behavior in these games, as well as defining good strategies to reduce it (in the same spirit of approximation and online algorithms' design), have stood as fundamental challenges. In this book, we show how these challenges can be fruitfully addressed via linear programming and duality theory. In particular, the presented results are two-fold:

Analysis (Part III): We measure the efficiency of selfish behavior in several classes of congestion games, and we show how this efficiency changes when considering different solution concepts, different types of latency functions (from linear and polynomial, to very general ones) and how much this efficiency eventually improves after strengthening the combinatorial properties of the considered games (e.g., when moving from general congestion games to load balancing games).

Design (Part IV): We resort to taxation mechanisms and Stackelberg strategies to improve the efficiency of almost all games we analyze in Part III. In most of the cases, we show that the performance of the mechanisms we propose is best possible within the considered category.

As a byproduct of our exposition, we also cover the analysis and design of efficient online algorithms for machine scheduling and load balancing problems (Awerbuch et al. [11]); indeed, the behavior of players entering a game sequentially simulates the execution of a proper *greedy* online algorithm.

From a technical point of view, the results we present here are based on the application of the *primal-dual method* (Bilò [22]). This is a powerful tool suited to prove good bounds (which are tight in most of the cases) on the performance guarantee of self-emerging solutions (such as pure Nash equilibria, one-round walks, and some of their generalizations) in several variants of congestion games. In particular, the method is based on the construction and analysis of a primal-dual pair of linear programs, where the dual is used to provide tight (or almost tight) upper bounds on the performance guarantee, and the primal is used to construct tight (or almost tight) lower bounds. In certain cases, the primal-dual method is equivalent to the *smoothness framework* of Roughgarden [158].

The primal-dual method, since its introduction, has been further explored and extended by the authors of this book, and has been successfully applied even outside the realm of congestion games. The results presented here give an overview on how to apply the method, by focusing on congestion games and some of their variants.

1.4 Organization of the Book

In the following, we summarize the content of all the subsequent book chapters:

Chapter 2: We provide the basic definitions and notation that will be adopted throughout the book and give a brief survey of the results related to the performance analysis of congestion games (e.g., in terms of price of anarchy, price of stability and competitive ratios of one-round walks).

Chapter 3: We describe in detail the primal-dual method. Most of the results presented in this book are based on the application of this method.

Chapter 4: We analyze the performance of both congestion games and load balancing games with very general latency functions. In particular, we show that the performance of load balancing games is as bad as that of general congestion games, under mild assumptions on the considered latency functions.

Chapter 5: We study ρ-uniform mixed equilibria in congestion games. This stability notion, which depends on an integer parameters $\rho \geq 1$, constitutes a sort of hybridization between the concepts of pure and mixed Nash equilibria. Moreover, it generalizes the notion of pure Nash equilibrium, as 1-uniform mixed equilibria coincide with pure Nash equilibria. We analyze the performance of ρ-uniform mixed equilibria in congestion games and quantify their price of anarchy and price of stability.

Chapter 6: We study the efficiency of congestion games with θ-similar strategies: a model of selfish routing in which the constant length of each route (i.e., the amount of travel time that is independent of the congestion) is at most θ times the length of any other one. We provide parametric tight bounds on the price of anarchy.

Chapter 7: We introduce a continuous variant of one-round walks for congestion games with general latency functions and provide closed-form tight bounds on the competitive ratio, holding even for load balancing games. Furthermore, we instantiate these formulas to the special case of games with polynomial latency functions.

Chapter 8: We introduce taxation mechanisms for congestion games and discuss some preliminary results.

Chapter 9: We study the efficiency of taxation in congestion games with polynomial latency functions. The presented results show a high beneficial effect which increases more than linearly with the degree of the latency functions; furthermore, the structure and analysis of the considered taxes exhibit interesting relationships with some well-studied polynomials in Combinatorics and Number Theory, such as the Touchard and the Geometric polynomials.

Chapter 10: The results provided in Chapter 9 are extended to the case of general latency functions, by providing closed-form bounds on their performance under taxation.

Chapter 11: We analyze the application of three Stackelberg strategies, namely Largest Latency First, Cover and Scale, to congestion games with affine latency functions, and we show upper and lower bounds on their worst-case price of anarchy.

At the end of each chapter, there is a dedicated section in which we discuss the implications of the presented results, and provide connections with the relevant state-of-the-art.

Some technical proofs are out of the scope of the book and are either omitted or sketched. To this aim, we often provide the main intuitions on how to prove a result, without considering all the technical (and possibly tedious) details. The complete proofs can be found in the research papers discussed in the "Concluding Remarks and Related Work" sections; furthermore, it becomes an interesting exercise for the reader to reconstruct the missing proofs.

1.5 Notation

In the following, we list most of the notation and symbols used in the book. The meaning of each symbol reported here is assumed to be fixed throughout all chapters, unless differently specified. The list is not exhaustive, as it does not contain symbols that are used in at most one chapter.

List of Symbols

Symbol	Meaning
$[k_1]_{k_2}$	$\{k_1, k_1+1, \ldots, k_2-1, k_2\}$
$[k]$	$[k]_1$
χ_A	indicator function ($\chi_A(x) = 1$ if $x \in A$, $\chi_A(x) = 0$ otherwise)
1^n	$\underbrace{(1,\ldots,1)}_{n \text{ times}}$
0^n	$\underbrace{(0,\ldots,0)}_{n \text{ times}}$
$\boldsymbol{x} \geq \boldsymbol{y}$	$x_i \geq y_i$ for any $i \in [n]$
$\boldsymbol{x} > \boldsymbol{y}$	$\boldsymbol{x} \geq \boldsymbol{y}$ and $x_i > y_i$ for some $i \in [n]$
$[a,b]$	interval $\{x \in \mathbb{R} : a \leq x \leq b\}$, $a, b \in \mathbb{R} \cup \{-\infty, \infty\}$
$(a,b]$	interval $\{x \in \mathbb{R} : a < x \leq b\}$, $a, b \in \mathbb{R} \cup \{-\infty, \infty\}$
$[a,b)$	interval $\{x \in \mathbb{R} : a \leq x < b\}$, $a, b \in \mathbb{R} \cup \{-\infty, \infty\}$
(a,b)	interval $\{x \in \mathbb{R} : a < x < b\}$, $a, b \in \mathbb{R} \cup \{-\infty, \infty\}$
2^E	set of all the subsets of set E
$f \circ g$	composition of functions f and g (i.e., $(f \circ g)(x) = f(g(x))$ for any x in the domain of f)
$\Theta(f(x))$	Theta asymptotic notation
$\Omega(f(x))$	Big-Omega asymptotic notation

$\omega(f(x))$	Little-Omega asymptotic notation
$O(f(x))$	Big-O asymptotic notation
$o(f(x))$	Little-O asymptotic notation
$\mathbb{P}[A]$	probability that event A occurs
$\mathbb{E}[X]$	expected value (of a random variable X)
$\mathbb{N}$	natural numbers
$\mathbb{Z}$	integer numbers
$\mathbb{Z}_{\geq}0$	non negative integer numbers
$\mathbb{R}$	real numbers
$\mathbb{R}_{\geq 0}$	non negative real-numbers
$\mathbb{R}_{>0}$	positive real numbers
$:=$	assignment operator
CG	generic congestion game (generally refers to atomic games)
LB	generic load balancing game (or load balancing instance with unrelated machines)
NCG	generic non-atomic game
N	set of players (resp. types of players) for atomic (resp. non-atomic) games
n	cardinality of N (or number of jobs in load balancing with unrelated machines)
i	generic player (or generic machine in load balancing with unrelated machines)
j	generic resource (or generic job in load balancing with unrelated machines)
E	set of resources
e	generic resource
w_i	weight of player i (it is different from symbol $w_i(\boldsymbol{\sigma})$)
$w_{i,j}$	weight of job j when assigned to machine i (in load balancing with unrelated machines)
$w_i(\boldsymbol{\sigma})$	load of machine i under assignment $\boldsymbol{\sigma}$ (in load balancing with unrelated machines)
Σ_i	set of strategies of player i
m_i	$\|\Sigma_i\|$ (i.e., number of strategies of player i)
$S_{i,j}$	jth strategy of Σ_i
r_i	total amount of players of type i (for non-atomic games only)
ℓ_e	generic latency function of resource e
f_e	f_e (or $\alpha_e f_e$, or $\alpha_e f_e + \beta_e$, with $\alpha_e, \beta_e \geq 0$) often denotes a generic latency function of resource e
S_i	generic strategy of player i
$\boldsymbol{\sigma}$ (or $\boldsymbol{s}$)	generic strategy profile $(\sigma_1, \ldots, \sigma_n)$ (or equilibrium, or strategy profile generated by a one-round walk)
σ_i (or s_i)	strategy of player i in strategy profile $\boldsymbol{\sigma}$ (or $\boldsymbol{s}$)

$\sigma_{i,S}$	total amount of players of type i selecting strategy S under profile $\boldsymbol{\sigma}$ (for non-atomic games only)
$\boldsymbol{\sigma}^*$	optimal strategy profile
σ_i^*	strategy of player i in strategy profile $\boldsymbol{\sigma}^*$
τ	$\boldsymbol{\tau} = (\boldsymbol{\sigma}^0, \boldsymbol{\sigma}^1, \ldots, \boldsymbol{\sigma}^n)$ denotes a generic approximate one-round walk starting from the empty state, or an online process
α_e	generic primal variable (that multiplies f_e)
γ, x	generic dual variables
$k_e(\boldsymbol{\sigma})$	weighted congestion of resource e in strategy profile $\boldsymbol{\sigma}$
k_e	congestion of resource e at the equilibrium (or in strategy profiles generated by one-round-walks)
o_e	congestion of resource e at the optimum
$n_e(\boldsymbol{\sigma})$	number of players selecting e in strategy profile $\boldsymbol{\sigma}$
n_e	$n_e(\boldsymbol{\sigma})$
n_e^*	$n_e(\boldsymbol{\sigma}^*)$
$c_i(\boldsymbol{\sigma})$ (or $cost_i(\boldsymbol{\sigma})$)	cost of player i in strategy profile $\boldsymbol{\sigma}$
$c_S(\boldsymbol{\sigma})$ (or $cost_S(\boldsymbol{\sigma})$)	cost of a player selecting strategy S in strategy profile $\boldsymbol{\sigma}$ (for non-atomic games only)
$\mathcal{G}$	generic class of congestion games
$\mathcal{C}$	generic class of latency functions
$\mathcal{C}(\mathcal{G})$	class of latency functions of congestion games in $\mathcal{G}$
f	generic latency function in $\mathcal{C}$
$\sum_{h=0}^{d} \alpha_{e,d} x^h$	generic polynomial latency function of maximum degree d
d (or p)	maximum degree of a polynomial latency
$\mathcal{P}(d)$	class of polynomial latencies of maximum degree d
$\mathsf{W}(\mathcal{C})$	weighted congestion games with latencies in $\mathcal{C}$
$\mathsf{U}(\mathcal{C})$	unweighted congestion games with latencies in $\mathcal{C}$
$\mathsf{N}(\mathcal{C})$	non-atomic congestion games with latencies in $\mathcal{C}$
$\mathsf{WLB}(\mathcal{C})$	weighted load balancing games with latencies in $\mathcal{C}$
$\mathsf{ULB}(\mathcal{C})$	unweighted load balancing games with latencies in $\mathcal{C}$
$\mathsf{NLB}(\mathcal{C})$	non-atomic load balancing games with latencies in $\mathcal{C}$
$\mathsf{WSLB}(\mathcal{C})$	symmetric weighted load balancing games with latencies in $\mathcal{C}$
$\mathsf{SUM}(\boldsymbol{\sigma})$	weighted or unweighted total latency of a strategy profile $\boldsymbol{\sigma}$
$\mathsf{WTL}(\boldsymbol{\sigma})$	weighted total latency of a strategy profile $\boldsymbol{\sigma}$
$\mathsf{TL}(\boldsymbol{\sigma})$	(unweighted) total latency of a strategy profile $\boldsymbol{\sigma}$
$\mathsf{TL}_p(\boldsymbol{\sigma})$	LP_p-norm of the total load of a strategy profile $\boldsymbol{\sigma}$
$\mathsf{WTL}_f(\boldsymbol{\sigma})$	f-weighted total load of a strategy profile $\boldsymbol{\sigma}$ (see Section 2.6)
$\mathsf{SF}(\boldsymbol{\sigma})$	generic social function applied to strategy profile $\boldsymbol{\sigma}$
$\mathsf{NE}_\epsilon(\mathsf{CG})$	set of ϵ-approximate pure Nash equilibria of game CG

$\mathsf{ORW}^s_\varepsilon(\mathsf{CG})$	set of strategy profiles generated by ε-approximate one-round walks involving selfish players, in game CG
$\mathsf{ORW}^c_\varepsilon(\mathsf{CG})$	set of strategy profiles generated by ε-approximate one-round walks involving cooperative players, in game CG
SC	generic solution concept ($\mathsf{SC} \in \{\mathsf{NE}_\varepsilon, \mathsf{ORW}^s_\varepsilon, \mathsf{ORW}^c_\varepsilon\}$)
$\mathsf{SC}(\mathsf{CG})$	set of strategy profiles of CG implementing solution concept SC
$\mathsf{PoA}_\varepsilon(\mathsf{CG})$	ε-approximate price of anarchy of congestion game CG
$\mathsf{PoA}_\varepsilon(\mathcal{G})$	ε-approximate price of anarchy of class $\mathcal{G}$
$\mathsf{CR}^s_\varepsilon(\mathsf{CG})$	competitive ratio of ε-approximate one-round walks involving selfish players of game CG
$\mathsf{CR}^s_\varepsilon(\mathcal{G})$	competitive ratio of ε-approximate one-round walks involving selfish players of class $\mathcal{G}$
$\mathsf{CR}^c_\varepsilon(\mathsf{CG})$	competitive ratio of ε-approximate one-round walks involving cooperative players of game CG
$\mathsf{CR}^c_\varepsilon(\mathcal{G})$	competitive ratio of ε-approximate one-round walks involving cooperative players of class $\mathcal{G}$
EM	generic efficiency metric ($\mathsf{EM} \in \{\mathsf{PoA}_\varepsilon, \mathsf{CR}^s_\varepsilon, \mathsf{CR}^c_\varepsilon\}$)
I	set of machines (in load balancing with unrelated machines)
J	set of jobs (in load balancing with unrelated machines)
$J_i(\boldsymbol{\sigma})$	set of jobs assigned to machine i in $\boldsymbol{\sigma}$ (in load balancing with unrelated machines)
(CG, δ)	generic congestion game with players subject to tax function δ
$\mathsf{SC}(\mathsf{CG}, \delta)$	set of strategy profiles of (CG, δ) implementing SC
$\delta_e(w, k_e)$	generic tax that a player of weight w pays to select a resource e with congestion k_e
$\delta_{\ell_e}(w, \mathbb{E}[k_e(\boldsymbol{y})], \ldots, k_e)$ (or $\delta_{\ell_e}(w, \boldsymbol{y}, \ldots, k_e)$, or $\delta(k_e, \mathbb{E}[k_e(\boldsymbol{y})], \ell_e)$)	tax function on a resource e with congestion k_e, based on mixed strategy profile $\boldsymbol{y}$ (the input clarifies the parameters which the tax is based on)
$\mathsf{TAX}_{\mathsf{MSP}}(\mathcal{G})$	set of taxes based on mixed strategy profiles, for a class of congestion games $\mathcal{G}$
$\delta_{\ell_e}(w_e, o_e, \ldots, k_e)$ (or $\delta(k_e, o_e, \ell_e)$)	tax function on a resource e with congestion k_e, based on optimal strategy profile $\boldsymbol{\sigma}^*$ (the input clarifies the parameters which the tax is based on)
$\mathsf{TAX}_{\mathsf{OPT}}(\mathcal{G})$	set of taxes based on optimal strategy profiles, for a class of congestion games $\mathcal{G}$

δ_h	tax function defined as δ_ℓ, but associated to the monomial x^h
$\hat{c}_i$	cost function of player i when she is charged a tax
$\mathsf{WTC}(\boldsymbol{\sigma})$	weighted total cost of $\boldsymbol{\sigma}$ when taxes are taken into account
$\mathsf{TC}(\boldsymbol{\sigma})$	(unweighted) total cost of $\boldsymbol{\sigma}$ when taxes are taken into account
$\mathcal{S}(i,j)$	(i,j)th Stirling Number of the Second Kind (see Section 8.2)
$\mathcal{B}_i(x)$	i-th Touchard (or Bell) polynomial (see Section 8.2)
$\mathcal{E}_i(x)$	i-th Geometric polynomial (see Section 8.2)
$\mathcal{Z}_i(x)$	another polynomial (see Section 8.2)
$\mathcal{R}_i(x)$	another polynomial (see Section 8.2)
$\mathcal{Q}_i(x)$	another polynomial (see Section 8.2)
$\Delta^i[x]\|_{x=k}$	ith forward finite difference (see Section 8.2)
$\mathcal{B}_\mathcal{G}(1+\varepsilon)$	see Chapter 10
$\mathcal{E}_\mathcal{G}(1+\varepsilon)$	see Chapter 10
$\mathcal{N}_\mathcal{G}(1+\varepsilon)$	see Chapter 10

Part II
Preliminaries

Chapter 2
The Performance of Congestion Games under Selfish and Greedy Dynamics: A Brief Survey

In this chapter, we provide the main notions and definitions that are helpful for understanding the other chapters of the book. Furthermore, we give an overview of the state-of-the-art concerning the characterization of the efficiency of outcomes generated by selfish and/or greedy dynamics in congestion games (and some of their variants). For the sake of brevity, the research work covered in some of the subsequent chapters is not mentioned here.

2.1 Preliminary Notation

We start with some preliminary notation that will be widely used throughout the book:

- Given two integers $0 \leq k_1 \leq k_2$, define $[k_2]_{k_1} := \{k_1, k_1+1, \ldots, k_2-1, k_2\}$ and $[k_2] := [k_2]_1$.
- Given two integers i, j, let $(i)_j$ denote the falling factorial $i(i-1)\cdots(i-j+1)$.
- For a set A, let $\chi_A := A \rightarrow \{0,1\}$ denote the *indicator function*, i.e., $\chi_A(x) = 1$ if $x \in A$, and $\chi_A(x) = 0$ otherwise.
- Vectors are generally denoted with bold letters (for instance, $\boldsymbol{x} = (x_1, \ldots, x_n)$).
- Let $\mathbf{1}^n$ and $\mathbf{0}^n$ denote the vectors $\underbrace{(1,\ldots,1)}_{n \text{ times}} \in \mathbb{R}^n_{\geq 0}$ and $\underbrace{(0,\ldots,0)}_{n \text{ times}} \in \mathbb{R}^n_{\geq 0}$, respectively.
- Given two vectors $\boldsymbol{x} = (x_1, \ldots, x_n)$ and $\boldsymbol{y} = (y_1, \ldots, y_n)$, notation $\boldsymbol{x} \geq \boldsymbol{y}$ means that $x_i \geq y_i$ for any $i \in [n]$; analogously, $\boldsymbol{x} > \boldsymbol{y}$ means that $\boldsymbol{x} \geq \boldsymbol{y}$ and $x_i > y_i$ for some $i \in [n]$.
- Given a set E, let 2^E denote the set of all subsets of E.
- $:=$ is the assignment operator, i.e., $A := B$ means that B is denoted as A; instead $A = B$ means that A is equal to B, not by definition. Anyway, when clear from the context, and with a little abuse of notation, we may use $=$ in place of $:=$ as assignment operator.

V. Bilò, C. Vinci, *Coping with Selfishness in Congestion Games*, Monographs in Theoretical Computer Science. An EATCS Series, https://doi.org/10.1007/978-3-031-30261-9_2

- In all definitions having a prefix "ε-approximate" or "ε-", when this prefix is either removed or replaced by the prefix "exact", we are implicitly assuming that $\varepsilon = 0$.

2.2 Atomic Congestion Games

A congestion game is called *atomic* when the set of players is finite, and the contribution of each player to the social welfare may be very significant. The following definition provides the general model of atomic congestion games, allowing also different players' weights.

Definition 2.1 (Fotakis et al. [91], Rosenthal [154]). A *weighted atomic congestion game* is a tuple

$$\mathsf{CG} = (N, E, (\ell_e)_{e\in E}, (w_i)_{i\in N}, (\Sigma_i)_{i\in N}),$$

where:
- $N := [n]$ is a finite set of $n \geq 2$ *players*;
- $E = [m]$ is a finite set of $m \geq 2$ *resources*;
- $\ell_e : \mathbb{R}_{>0} \to \mathbb{R}_{>0}$ is the *latency function* of resource $e \in E$, that is assumed to be non-decreasing and positive[1];
- for each $i \in N$, $w_i > 0$ is the *weight* of player $i \in N$;
- $\Sigma_i = \{S_{i,j} : j \in [m_i]\} \subseteq 2^E \setminus \{\emptyset\}$ is the set of m_i *strategies* available to player $i \in N$.

We extend the domain of the latency functions ℓ_e to $x = 0$, by setting $\ell_e(0) := \lim_{x\to 0^+} \ell_e(x)$. We speak of *weighted* games/players when players have arbitrary weights and of *unweighted* games/players when $w_i = 1$ for each $i \in N$.

Players' Strategy Space

Interesting special cases of congestion games are obtained by restricting the combinatorics of the players' strategic space.

- A congestion game is *symmetric* (Christodoulou and Koutsoupias [59], Fabrikant et al. [83]) if $\Sigma_i = \Sigma_j := \Sigma$ for each $i, j \in N$, i.e., if all players share the same strategy space, denoted by Σ.
- A *load balancing game* (or *singleton congestion game*) (Ackermann et al. [3], Caragiannis et al. [54], Nisan et al. [134], Suri et al. [170]) is a congestion game

[1] All the results obtained in this book can be easily generalized to the case in which some latency function ℓ_e verifies $\ell_e(x) = 0$ for some $x > 0$, but this requires further (tedious) case analysis to cope with non-definite ratios of type $c/0$.

in which, for each $i \in N$ and $S \in \Sigma_i$, $|S| = 1$, that is, all players' strategies are singleton sets; furthermore, we speak of *symmetric load balancing games* (or *parallel-link games*) if the considered load balancing games are also symmetric, that is, all players can select all singletons.

- In *network congestion games* (Fabrikant et al. [83], Nisan et al. [134], Roughgarden [155]), the strategies of each player $i \in N$ are all the paths connecting a source node s_i with a destination node t_i in a given network. If $s_i = s$ for any player $i \in N$, we speak of *single-source network congestion games*; if $s_i = s$ and $t_i = t$ for any player $i \in N$, we speak of *symmetric network congestion games*; finally, if $s_i = s$ and $t_i = t$ for any player $i \in N$ and all (s,t)-paths do not share any node, except for s and t, we speak of *path-disjoint network congestion games*. One can easily show that load balancing (resp. symmetric load balancing) games can be modeled as particular instances of single-source (resp. path-disjoint) network congestion games.
- In *matroid congestion games* (Ackermann et al. [3, 4]), the strategy set of every player is given by the set of bases of a matroid defined over the set of available resources; in *k-uniform matroid congestion games* (de Jong and Uetz [80], de Jong et al. [81]), each player can select any subset of cardinality k from a prescribed player-specific set of resources. We observe that symmetric load balancing games can be equivalently defined as 1-uniform matroid congestion games.

In the remainder of the book, we mainly focus on general congestion games or load balancing games. Given a class $\mathcal{C}$ of latency functions, we have that

- $\mathsf{W}(\mathcal{C})$ denotes the class of weighted congestion games,
- $\mathsf{U}(\mathcal{C})$ denotes the class of unweighted congestion games,
- $\mathsf{WLB}(\mathcal{C})$ denotes the class of weighted load balancing games,
- $\mathsf{WSLB}(\mathcal{C})$ denotes the class of weighted symmetric load balancing games,
- $\mathsf{ULB}(\mathcal{C})$ denotes the class of unweighted load balancing games,

all having latency functions in class $\mathcal{C}$.

Latency Functions

In the following, we list several classes of latency functions that we treat in this book.

- A congestion game has *polynomial latency functions* of maximum degree $d \in \mathbb{N}$ when, for each $e \in E$, $\ell_e(x) = \sum_{h \in [d]_0} \alpha_{e,h} \cdot x^h$, with $\alpha_{e,h} \geq 0$ for each $h \in [d]_0$. When $d = 0$ we speak of *constant latencies*, when $d = 1$, we speak of *affine latencies*, and if $\ell_e(x) = \alpha_{e,1} \cdot x$ we speak of *linear latencies*. Let $\mathcal{P}(d)$ denote the class of polynomial latencies of maximum degree d.
- A latency function f is *semi-convex* (resp. *strictly semi-convex*) if function $xf(x)$ is convex (resp. strictly convex); we observe that polynomial latency functions are semi-convex.

- A latency function f is *unbounded* if $\lim_{x\to\infty} f(x) = \infty$.
- A congestion game has *identical resources* when all resources have the same latency function.
- Given a class $\mathcal{G}$ of congestion games, let $\mathcal{C}(\mathcal{G})$ denote the class of latency functions of congestion games belonging to $\mathcal{G}$.
- A class $\mathcal{C}$ of latency function is *closed under ordinate scaling* (resp. *abscissa scaling*) if, for any function $f \in \mathcal{C}$ and $\alpha \geq 0$, the function g such that $g(x) = \alpha f(x)$ (resp. $g(x) = f(\alpha x)$) belongs to $\mathcal{C}$.

Strategy Profiles and Cost Functions

A *(pure) strategy profile* is an n-tuple of strategies $\boldsymbol{\sigma} = (\sigma_1, \ldots, \sigma_n)$, that is, a state of the game in which each player $i \in N$ is adopting strategy $\sigma_i \in \Sigma_i$, so that $\boldsymbol{\Sigma} := \times_{i\in N} \Sigma_i$ denotes the set of strategy profiles which can be realized in CG. For a strategy profile $\boldsymbol{\sigma}$, the *congestion* of resource $e \in E$ in $\boldsymbol{\sigma}$, denoted as

$$k_e(\boldsymbol{\sigma}) := \sum_{i\in N: e\in\sigma_i} w_i,$$

is the total weight of the players using resource e in $\boldsymbol{\sigma}$, while the number of users of resource e in $\boldsymbol{\sigma}$ is denoted as $n_e(\boldsymbol{\sigma}) := |\{i \in N : e \in \sigma_i\}|$. The *cost* of player i in $\boldsymbol{\sigma}$ is defined as

$$c_i(\boldsymbol{\sigma}) = \sum_{e\in\sigma_i} \ell_e(k_e(\boldsymbol{\sigma}))$$

(equivalently denoted as $cost_i(\boldsymbol{\sigma})$) and each player aims at minimizing it. For the sake of conciseness, when the strategy profile $\boldsymbol{\sigma}$ is clear from the context, we will often write k_e and n_e in place of $k_e(\boldsymbol{\sigma})$ and $n_e(\boldsymbol{\sigma})$.

We also treat the notion of *mixed strategy profiles* (Nash [133]), which constitute a probabilistic extension of pure ones. Let

$$\Delta_i = \left\{ \boldsymbol{y}_i = (y_{i,j})_{j\in[m_i]} \in \mathbb{R}^{m_i}_{\geq 0} : \sum_{j\in m_i} y_{i,j} = 1 \right\}$$

be the set of probability distributions defined over Σ_i, so that $y_{i,j}$ denotes the probability for player i to choose strategy $S_{i,j}$. A *mixed strategy profile* $\boldsymbol{y} = (\boldsymbol{y}_1, \boldsymbol{y}_2, \ldots, \boldsymbol{y}_n)$ is a state in which each player $i \in N$ picks her strategy according to probability distribution $\boldsymbol{y}_i \in \Delta_i$, independently from the choices of the other players. Let $\boldsymbol{\Delta} = \times_{i\in N} \Delta_i$ be the set of all mixed strategy profiles. Given $\boldsymbol{y} \in \boldsymbol{\Delta}$, denote with

$$\mathbb{E}[k_e(\boldsymbol{y})] = \sum_{i\in N} \sum_{j\in[m_i]: e\in S_{i,j}} w_i y_{i,j}$$

the expected congestion of resource e according to $\boldsymbol{y}$, and with

$$\mathbb{E}[n_e(\mathbf{y})] = \sum_{i\in N}\sum_{j\in[m_i]:e\in S_{i,j}} y_{i,j}$$

the expected number of users of e according to $\mathbf{y}$. Note that each strategy profile is also a mixed one and that, whenever a mixed strategy profile $\mathbf{y}$ coincides with the strategy profile $\boldsymbol{\sigma}$, we have $\mathbb{E}[k_e(\mathbf{y})] = k_e(\boldsymbol{\sigma})$ and $\mathbb{E}[n_e(\mathbf{y})] = n_e(\boldsymbol{\sigma})$. The cost of player i in a mixed profile $\mathbf{y}$ is defined as

$$c_i(\mathbf{y}) := \mathbb{E}_{\boldsymbol{\sigma}\sim\mathbf{y}}[c_i(\boldsymbol{\sigma})],$$

i.e., it is the expected cost of player i with respect to strategy profiles distributed according to $\mathbf{y}$.

The more general notion of *randomized strategy profiles* (Aumann [10], Moulin and Vial [131]) can be obtained by relaxing the independence between the players' random choices, and are defined as arbitrary probability distributions over pure strategy profiles.

Social Functions

The quality of a strategy profile in congestion games is measured by means of a *social function*. The following two social functions are usually used in weighted congestion games: the *total latency* (or *sum of all latencies*), defined as

$$\mathsf{TL}(\boldsymbol{\sigma}) = \sum_{i\in N} c_i(\boldsymbol{\sigma}) = \sum_{e\in E} n_e(\boldsymbol{\sigma})\ell_e(k_e(\boldsymbol{\sigma}))$$

and the *weighted total latency* (or *weighted sum of all latencies*), defined as

$$\mathsf{WTL}(\boldsymbol{\sigma}) = \sum_{i\in N} w_i c_i(\boldsymbol{\sigma}) = \sum_{e\in E} k_e(\boldsymbol{\sigma})\ell_e(k_e(\boldsymbol{\sigma})).$$

Clearly, by definition, these two measures coincide in unweighted games, and for the sake of simplicity, we either adopt both WTL and TL as acronyms for the total latency in unweighted games, or we adopt the acronym SUM.

Moreover, observe that, for polynomial latencies, $\mathsf{TL}(\boldsymbol{\sigma}) = \sum_{e\in E}\sum_{h\in[d]_0}\alpha_{e,h}n_e k_e^h$ and $\mathsf{WTL}(\boldsymbol{\sigma}) = \sum_{e\in E}\sum_{h\in[d]_0}\alpha_{e,h}k_e^{h+1}$. Other social functions are the *maximum cost*, defined as

$$\mathsf{MAX}(\boldsymbol{\sigma}) = \max_{i\in N} c_i(\boldsymbol{\sigma}),$$

and the *L_p-norm of the total latency*, defined as

$$\mathsf{TL}_p(\boldsymbol{\sigma}) = \sqrt[p]{\sum_{e\in E}\ell_e(k_e(\boldsymbol{\sigma}))^p}.$$

The results presented in this book are mainly related to social functions TL and WTL. In particular, when the social function is not specified, we implicitly refer to the total latency when the game is unweighted, and to the weighted total latency when the game is weighted.

For a fixed social function SF, a *social optimum* is a strategy profile $\boldsymbol{\sigma}^*$ minimizing SF. For the sake of conciseness, once a particular social optimum has been fixed, we will often write o_e and n_e^* to denote the values $k_e(\boldsymbol{\sigma}^*)$ and $n_e(\boldsymbol{\sigma}^*)$, respectively.

2.3 Solution Concepts

In unweighted congestion games, selfish behavior always leads to stable outcomes, called *pure Nash equilibria* (Nash [133]), in which each player cannot improve her utility by unilaterally deviating from her strategy. Besides pure Nash equilibria, in the setting of atomic congestion games, the notion of one-round walks starting from the empty state has been widely investigated due to its simplicity and effectiveness. This concept assumes that, starting from the situation in which no strategy has been specified yet, the players are processed sequentially and, at each iteration, the selected player irrevocably chooses her strategy so as to minimize her cost based on the choices of the previous ones. In the following, we formally describe these two solution concepts.

Pure Nash Equilibrium and Generalizations

Fix a strategy profile $\boldsymbol{\sigma}$ and a player $i \in N$. We denote with $\boldsymbol{\sigma}_{-i}$ the restriction of $\boldsymbol{\sigma}$ to all players other than i; moreover, for a strategy $S \in \Sigma_i$, we denote with $(\boldsymbol{\sigma}_{-i}, S)$ the strategy profile obtained from $\boldsymbol{\sigma}$ when player i changes her strategy from σ_i to S, while the strategies of all the other players are kept fixed.

Definition 2.2 (Nash [133]). A *pure Nash equilibrium* is a strategy profile $\boldsymbol{\sigma}$ such that, for any player $i \in N$ and strategy $S \in \Sigma_i$, $c_i(\boldsymbol{\sigma}) \leq c_i(\boldsymbol{\sigma}_{-i}, S)$, that is, an outcome of the game in which no player can improve her situation by unilaterally deviating to another strategy.

By relaxing this requirement, one obtains the notion of ε-approximate pure Nash equilibrium as follows.

Definition 2.3. For any $\varepsilon \geq 0$, an *ε-approximate pure Nash equilibrium* is a strategy profile $\boldsymbol{\sigma}$ such that, for any player $i \in N$ and strategy $S \in \Sigma_i$, $c_i(\boldsymbol{\sigma}) \leq (1+\varepsilon)c_i(\boldsymbol{\sigma}_{-i}, S)$.

We denote by $\mathsf{NE}_\varepsilon(\mathsf{CG})$ the set of ε-approximate pure Nash equilibria of a congestion game CG.

A canonical way to show the existence of (approximate) pure Nash equilibria is by means of a potential function (Monderer and Shapley [128], Rosenthal [154]). More precisely, given a (congestion) game SG, a function $\Phi : \boldsymbol{\Sigma} \to \mathbb{R}_{\geq 0}$ is called

- an *ordinal potential* if, for any strategy profile $\boldsymbol{\sigma} \in \boldsymbol{\Sigma}$, player $i \in N$ and strategy $S \in \Sigma_i$, it holds that $sgn(\Phi(\boldsymbol{\sigma}) - \Phi(\boldsymbol{\sigma}_{-i}, S)) = sgn(c_i(\boldsymbol{\sigma}) - c_i(\boldsymbol{\sigma}_{-i}, S))$;
- a *weighted potential* if, for any player $i \in N$, there exists a weight $v_i > 0$ such that for any strategy profile $\boldsymbol{\sigma} \in \boldsymbol{\Sigma}$ and strategy $S \in \Sigma_i$, it holds that $\Phi(\boldsymbol{\sigma}) - \Phi(\boldsymbol{\sigma}_{-i}, S) = v_i(c_i(\boldsymbol{\sigma}) - c_i(\boldsymbol{\sigma}_{-i}, S))$;
- an *exact potential* if for any strategy profile $\boldsymbol{\sigma} \in \boldsymbol{\Sigma}$, player $i \in N$ and strategy $S \in \Sigma_i$, it holds that $\Phi(\boldsymbol{\sigma}) - \Phi(\boldsymbol{\sigma}_{-i}, S) = c_i(\boldsymbol{\sigma}) - c_i(\boldsymbol{\sigma}_{-i}, S)$.

Clearly, by definition, exact potentials are special cases of weighted potentials, which, in turn, are special cases of ordinal ones. The existence of an ordinal potential function implies that any dynamics in which each player decreases her cost by deviating to other strategies converges to a pure Nash equilibrium after a finite number of steps, since the set of all strategy profiles is finite.

Rosenthal [154] shows that any unweighted congestion game admits an exact potential function, defined as

$$\Phi(\boldsymbol{\sigma}) = \sum_{e \in E} \sum_{i=1}^{k_e(\boldsymbol{\sigma})} \ell_e(i).$$

Conversely, Monderer and Shapley [128] show that any game admitting an exact potential function is equivalent to a congestion game. The works of Fotakis et al. [91], Harks [100], Panagopoulou and Spirakis [142], and Harks and Klimm [101] have shown that pure Nash equilibria are always guaranteed to exist in weighted congestion games if and only if the latency functions are either affine or exponential.

As to approximate equilibria, Caragiannis et al. [55] show that weighted congestion games with polynomial latencies of maximum degree d always admit an $O(d!)$-approximate pure Nash equilibrium; this result has been further improved by Caragiannis and Fanelli [51], who show the existence of $O(d)$-approximate pure Nash equilibria. In contrast, Christodoulou et al. [66] show that $\Omega(\sqrt{d})$-approximate equilibria might not exist for certain weighted congestion games with polynomial latencies of maximum degree d.

Several generalizations of (approximate) pure Nash equilibria have been proposed in the literature (Aumann [10], Moulin and Vial [131], Nash [133]). Here, we present their approximate variants.

Definition 2.4. An *ε-approximate mixed Nash equilibrium* is a mixed strategy profile $\boldsymbol{y} \in \boldsymbol{\Delta}$ such that, for any player $i \in N$ and strategy $S \in \Sigma_i$,

$$\mathbb{E}_{\boldsymbol{\sigma} \sim \boldsymbol{y}}[c_i(\boldsymbol{\sigma})] \leq (1+\varepsilon)\mathbb{E}_{\boldsymbol{\sigma} \sim \boldsymbol{y}}[c_i(\boldsymbol{\sigma}_{-i}, S)].$$

Definition 2.5. An *ε-approximate correlated equilibrium* is a probability distribution $\boldsymbol{p}$ defined over the set of strategy profiles $\boldsymbol{\Sigma}$ (that is, a randomized strategy profile) such that, for any player $i \in N$ and pair of strategies $S, S' \in \Sigma_i$,

$$\mathbb{E}_{\boldsymbol{\sigma}\sim\boldsymbol{p}}[c_i(\boldsymbol{\sigma})|\sigma_i = S'] \leq (1+\varepsilon)\mathbb{E}_{\boldsymbol{\sigma}\sim\boldsymbol{p}}[c_i(\boldsymbol{\sigma}_{-i}, S)|\sigma_i = S'].$$

Definition 2.6. an *ε-approximate coarse correlated equilibrium* is a probability distribution $\boldsymbol{p}$ defined over $\boldsymbol{\Sigma}$ such that, for any player $i \in N$ and strategy $S \in \Sigma_i$,

$$\mathbb{E}_{\boldsymbol{\sigma}\sim\boldsymbol{p}}[c_i(\boldsymbol{\sigma})] \leq (1+\varepsilon)\mathbb{E}_{\boldsymbol{\sigma}\sim\boldsymbol{p}}[c_i(\boldsymbol{\sigma}_{-i}, S)].$$

We can easily observe that every ε-pure Nash equilibrium is an ε-approximate mixed Nash equilibrium, every ε-approximate mixed Nash equilibrium is an ε-approximate correlated equilibrium, and every ε-approximate correlated equilibrium is an ε-approximate coarse correlated equilibrium.

By Nash's celebrated theorem (Nash [133]), mixed/correlated/coarse correlated Nash equilibria are always guaranteed to exist in weighted congestion games for any $\varepsilon \geq 0$. Moreover, correlated equilibria (and so also coarse correlated ones) can be efficiently computed, as shown by Papadimitriou and Roughgarden [144].

We have previously stated that pure Nash equilibria are always guaranteed to exist for subsets of weighted congestion games by resorting to a potential function argument. However, the problem of finding a pure Nash equilibrium in these games is computationally hard. Indeed, Fabrikant et al. [83] show that computing a pure Nash equilibrium is a PLS-complete problem (and this holds even for unweighted congestion games with affine latencies, as shown by Ackermann et al. [3]). The problem becomes polynomial when restricting the players' strategic space (e.g., to single-source network congestion games (Fabrikant et al. [83]), or to matroid congestion games (Ackermann et al. [3])).

For such a reason, less demanding deterministic solution concepts have been investigated in the literature, such as the notion of one-round walk introduced by Mirrokni and Vetta [127] which we explain in the following.

One-round Walk Starting From the Empty State

The Case of Selfish Players

For any $\varepsilon \geq 0$, a strategy $S^* \in \Sigma_i$ is an *ε-approximate best-response* for player i in $\boldsymbol{\sigma}_{-i}$ if, for each $S \in \Sigma_i$, $c_i(\boldsymbol{\sigma}_{-i}, S^*) \leq (1+\varepsilon)c_i(\boldsymbol{\sigma}_{-i}, S)$. With a little abuse of notation, we extend the set $\boldsymbol{\Sigma}$ of strategy profiles so as to include also the cases in which some players have not chosen their strategies yet; in particular, we denote with $\boldsymbol{\emptyset}$ the empty state, that is, the strategy profile in which none of the players has chosen a strategy.

Definition 2.7 (Mirrokni and Vetta [127]). An *ε-approximate one-round walk starting from the empty state (involving selfish players)* is an $(n+1)$-tuple of strategy profiles

$$\boldsymbol{\tau} = (\boldsymbol{\sigma}^0, \boldsymbol{\sigma}^1, \boldsymbol{\sigma}^2, \ldots, \boldsymbol{\sigma}^n),$$

such that $\boldsymbol{\sigma}^0 = \boldsymbol{\emptyset}$ and, for each $i \in N$, $\boldsymbol{\sigma}^i = (\boldsymbol{\sigma}^{i-1}_{-i}, S^*)$, where S^* is an ε-approximate best-response for player i in $\boldsymbol{\sigma}^{i-1}_{-i}$. The strategy profile $\boldsymbol{\sigma}^n$ is the strategy profile generated by $\boldsymbol{\tau}$.

The above definition is an approximate variant of the solution concept introduced by Mirrokni and Vetta [127]. The strategy profile generated by $\boldsymbol{\tau}$ can be seen as the result of the process in which players enter the game sequentially and the ith player performs an irrevocable choice by approximately best-responding to the choices of the first $i-1$ players.

To highlight the connection between the concept of one-round walk and the typical concepts of equilibrium in game theory, the notion of one-round walk has been equivalently defined as *myopic best-response equilibrium* (see, for instance, Heydenreich et al. [103]): indeed, a one round-walk can be seen as a process in which all players, in turn, select their best strategy without taking into account the behavior of the players which have not entered the game yet, i.e. players are myopic. A similar behavior in which each player entering the game is not myopic, but chooses her strategy taking also into account the strategic behavior of the players which have not entered the game yet, is captured by the well-known concept of subgame perfect equilibrium (Osborne [136]), whose performance has been studied in several works (see, for instance, Bilò et al. [33], Correa et al. [76], de Jong and Uetz [80], Paes Leme et al. [141]).

The Case of Cooperative Players

In Definition 2.7, if S^* is a strategy such that $\mathsf{WTL}(\boldsymbol{\sigma}^{i-1}_{-i}, S^*) - \mathsf{WTL}(\boldsymbol{\sigma}^{i-1}) \leq (1+\varepsilon)(\mathsf{WTL}(\boldsymbol{\sigma}^{i-1}_{-i}, S) - \mathsf{WTL}(\boldsymbol{\sigma}^{i-1}))$ for any $S \in \Sigma_i$, then $\boldsymbol{\tau}$ is an *ε-approximate one-round walk starting from the empty state involving cooperative players*[2]. This cooperative version of one-round walks, despite being not meaningful when assuming selfish players, is significant in the context of online algorithms, since the greedy-algorithm obtained by simulating a one-round walk has been used to solve efficiently the *online load balancing problem* (Awerbuch et al. [11], Caragiannis [48], Caragiannis et al. [54]) (for further details, see Section 2.6).

We denote by $\mathsf{ORW}^s_\varepsilon(\mathsf{CG})$ (resp. $\mathsf{ORW}^c_\varepsilon(\mathsf{CG})$) the set of strategy profiles which can be constructed by an ε-approximate one-round walk involving selfish (resp. cooperative) players in a congestion game CG.

[2] An alternative definition for the concept of ε-approximate one-round walk involving cooperative players can be obtained by assuming that the ith player aims at minimizing the value of the social function, i.e., $\mathsf{WTL}(\boldsymbol{\sigma}^i) \leq (1+\varepsilon)\min_{S\in\Sigma_i} \mathsf{WTL}(\boldsymbol{\sigma}^{i-1}_{-i}, S)$. For $\varepsilon = 0$ the two definitions are equivalent; for $\varepsilon > 0$, the proof arguments that we are going to use in this book can be easily adapted/modified to get similar results holding under this alternative definition as well.

2.4 Efficiency Metrics

In general, all solution concepts described above do not yield an optimal social welfare, and several metrics have been proposed to measure the inefficiency of these outcomes. All the metrics we consider here vary in the range $[1,\infty]$, in such a way that, the higher their value, the more inefficient the outcome (with respect to the considered solution concept and social function). In particular, if the adopted metric gets value 1, the outcomes of the game are optimal (in terms of social welfare).

Price of Anarchy and Price of Stability

The most important metrics used to measure the inefficiency of Nash equilibria are the *price of anarchy* (Koutsoupias and Papadimitriou [113]) and the *price of stability* (Anshelevich et al. [8]). We give a formal definition for these metrics, holding for general ε-approximate pure Nash equilibria.

Definition 2.8 (Koutsoupias and Papadimitriou [113]). The *ε-approximate pure price of anarchy* of a weighted congestion game CG with respect to the social function $\mathsf{SF} \in \{\mathsf{TL},\mathsf{WTL}\}$ is defined as

$$\mathsf{PoA}_\varepsilon(\mathsf{CG}) := \max_{\boldsymbol{\sigma}\in \mathsf{NE}_\varepsilon(\mathsf{CG})} \frac{\mathsf{SF}(\boldsymbol{\sigma})}{\mathsf{SF}(\boldsymbol{\sigma}^*)}, \tag{2.1}$$

where $\boldsymbol{\sigma}^*$ is a social optimum for CG with respect to the social function SF. Given a class of congestion games $\mathcal{G}$, the *ε-approximate pure price of anarchy* of class $\mathcal{G}$ is defined as $\mathsf{PoA}(\mathcal{G}) := \sup_{\mathsf{CG}\in\mathcal{G}} \mathsf{PoA}(\mathsf{CG})$.

Definition 2.9 (Anshelevich et al. [8]). The *ε-approximate pure price of stability* of a weighted congestion game CG with respect to the social function $\mathsf{SF} \in \{\mathsf{TL},\mathsf{WTL}\}$ is defined as

$$\mathsf{PoS}_\varepsilon(\mathsf{CG}) := \min_{\boldsymbol{\sigma}\in \mathsf{NE}_\varepsilon(\mathsf{CG})} \frac{\mathsf{SF}(\boldsymbol{\sigma})}{\mathsf{SF}(\boldsymbol{\sigma}^*)}, \tag{2.2}$$

where $\boldsymbol{\sigma}^*$ is a social optimum for CG with respect to the social function SF. Given a class of congestion games $\mathcal{G}$, the *ε-approximate pure price of stability* of class $\mathcal{G}$ is defined as $\mathsf{PoS}(\mathcal{G}) := \sup_{\mathsf{CG}\in\mathcal{G}} \mathsf{PoS}(\mathsf{CG})$.

By letting $\boldsymbol{\sigma}$ varying among the set of ε-approximate mixed Nash equilibria, ε-approximate correlated equilibria and ε-approximate coarse correlated equilibria, one respectively obtains the *ε-approximate mixed price of anarchy*, the *ε-approximate correlated price of anarchy* and the *ε-approximate coarse correlated price of anarchy*, with respect to social function SF. Analogously, one can define the *ε-approximate mixed/correlated/coarse correlated price of stability*.

Informally, the price of anarchy of a game is defined as the worst-case ratio between the social value of an equilibrium and the optimal social value, and measures the impact of selfish behavior when players are completely uncoordinated. Instead, the price of stability is defined as the best-case ratio between the above quantities, and then assumes that there is a partial coordination among the players, who are able to reach the best equilibrium configuration.

In all the chapters of this book, we assume without loss of generality that each considered class of latency functions contains at least a non-constant latency function.[3]

Competitive Ratio of One-round Walks

In the same spirit of online optimization, we adopt the *competitive ratio* as a benchmark measure for the performance of strategy profiles generated by one-round walks.

Definition 2.10. The *competitive ratio* of ε-approximate one-round walks generated by selfish (resp. cooperative) players of a weighted congestion game CG with respect to the social function $\mathsf{SF} \in \{\mathsf{TL}, \mathsf{WTL}\}$ is defined as

$$\mathsf{CR}^x_\varepsilon(\mathsf{CG}) = \max_{\boldsymbol{\sigma} \in \mathsf{ORW}^x_\varepsilon(\mathsf{CG})} \frac{\mathsf{SF}(\boldsymbol{\sigma})}{\mathsf{SF}(\boldsymbol{\sigma}^*)}, \tag{2.3}$$

where $x = s$ for selfish players (resp. $x = c$ for cooperative players), and $\boldsymbol{\sigma}^*$ is a social optimum for CG with respect to the social function SF. Given a class of congestion games $\mathcal{G}$, the *competitive ratio* of ε-approximate one-round walks of class $\mathcal{G}$ is defined as $\mathsf{CR}^x_\varepsilon(\mathcal{G}) = \sup_{\mathsf{CG} \in \mathcal{G}} \mathsf{CR}^x_\varepsilon(\mathsf{CG})$.

Similarly to the price of anarchy and the price of stability, the competitive ratio measures the inefficiency of the outcomes generated by players who greedily select their approximate best-responses.

2.5 Non-atomic Congestion Games

The counterpart of the class of atomic games is that of *non-atomic congestion games* (Beckmann et al. [17], Pigou [150], Wardrop [178]): these games constitute a good approximation for atomic congestion games when players become infinitely many and the contribution of each player to the social welfare becomes infinitesimally small.

---

[3] Indeed, if it were not the case (i.e., if all the latency functions were constant) there would be no interactions between players, and each player would be trivially able to minimize her cost up to a factor $(1+\varepsilon)$, independently from the choices of the others. This results in a social value equal to at most $(1+\varepsilon)$ times the optimal one, and this upper bound is trivially tight.

Definition 2.11 (Beckmann et al. [17], Pigou [150], Wardrop [178]). A *non-atomic congestion game* is a tuple

$$\mathsf{NCG} = (N, E, (\ell_e)_{e \in E}, (r_i)_{i \in N}, (\Sigma_i)_{i \in N}),$$

where

- $N := [n]$ is a finite set of $n \geq 1$ *types of players*, and $E := [m]$ is a finite set of $m \geq 2$ *resources*;
- $\ell_e : \mathbb{R}_{>0} \to \mathbb{R}_{>0}$ is the *latency function* of resource $e \in E$, which is assumed to be non-decreasing, positive, and continuous[4] (again, we set $\ell_e(0) := \lim_{x \to 0^+} \ell_e(x)$);
- $r_i \in \mathbb{R}_{\geq 0}$ is the *amount of players* of type $i \in N$;
- $\Sigma_i = \{S_{i,1}, \ldots, S_{i,m_i}\} \subseteq 2^E \setminus \{\emptyset\}$ is the set of *strategies* of a player of type i, that is, each player of type i has $m_i \geq 1$ possible choices.

As in the atomic case, we can define the concepts of strategy profile, cost function, and total latency.

A *(pure) strategy profile* is a tuple $\boldsymbol{\sigma} := (\sigma_{i,S})_{i \in N, S \in \Sigma_i}$ with $\sum_{S \in \Sigma_i} \sigma_{i,S} = r_i$ for any $i \in N$, that is a state of the game where $\sigma_{i,S} \geq 0$ is the total amount of players of type i selecting strategy S for any $i \in N$ and $S \in \Sigma_i$. Given a strategy profile $\boldsymbol{\sigma}$,

$$k_e(\boldsymbol{\sigma}) := \sum_{i \in N, S \in \Sigma_i : e \in S} \sigma_{i,S}$$

is the *congestion* of e in $\boldsymbol{\sigma}$, i.e., the total amount of players selecting e in $\boldsymbol{\sigma}$, and given a strategy S,

$$c_S(\boldsymbol{\sigma}) := \sum_{e \in S} \ell_e(k_e(\boldsymbol{\sigma}))$$

is the *cost* of players selecting S in $\boldsymbol{\sigma}$. A social function usually used as a measure of the quality of a strategy profile in non-atomic congestion games is the *total latency*, defined as

$$\mathsf{TL}(\boldsymbol{\sigma}) = \sum_{i \in N, S \in \Sigma_i} \sigma_{i,S} \cdot c_S(\boldsymbol{\sigma}) = \sum_{e \in E} k_e(\boldsymbol{\sigma}) \cdot \ell_e(k_e(\boldsymbol{\sigma})).$$

All the other definitions given in Section 2.2 for atomic congestion games (e.g., load balancing games, games with polynomial latency functions), naturally extend to non-atomic ones. Given a class $\mathcal{C}$ of latency functions, let $\mathsf{N}(\mathcal{C})$ be the class of non-atomic congestion games having latency functions in $\mathcal{C}$, and let $\mathsf{NLB}(\mathcal{C}) \subset (\mathcal{C})$ be the related subclass of load balancing games.

As in the atomic case, pure Nash equilibria are considered as synonymous of stability in non-atomic congestion games. In the following definition, we consider the approximate variant of pure Nash equilibria.

[4] The property of continuity is well-motivated by most of the real-life scenarios modeled by non-atomic congestion games. Anyway, the results provided in this book hold even under the weaker assumption of right-continuity.

Definition 2.12 (Nash [133], Wardrop [178]). An *ε-approximate pure Nash equilibrium* is a strategy profile $\boldsymbol{\sigma}$ such that, for any player $i \in N$, strategy $S \in \Sigma_i$ such that $\sigma_{i,S} > 0$ and strategy S', $c_S(\boldsymbol{\sigma}) \leq c_{S'}(\boldsymbol{\sigma})$ holds, that is, an outcome of the game in which no player can improve her situation by unilaterally deviating to another strategy.

In the literature of selfish routing, exact pure Nash equilibria (in non-atomic games) are often called *Wardrop equilibria* or *equilibrium flows*. As for atomic games, non-atomic congestion games always admit an exact pure Nash equilibrium, but in the non-atomic case, (exact) pure Nash equilibria are unique up to strategy profiles having the same value of the latency functions for any resource. Indeed, Beckmann et al. [17] prove that a strategy profile $\boldsymbol{\sigma}$ is a pure Nash equilibrium if and only if it is a local minimum of the potential function

$$\Phi(\boldsymbol{\sigma}) := \sum_{e \in E} \int_0^{k_e(\boldsymbol{\sigma})} \ell_e(t)dt.$$

Since $\Phi(\boldsymbol{\sigma})$ is convex over a compact and convex domain, it admits a global minimum, thus at least a Nash equilibrium exists. Furthermore, because of convexity, all local minima are global minima, and the value of the latency function of each resource does not depend on the particular equilibrium.

The notions of ε-approximate prices of anarchy and stability given in Section 2.4 naturally extend to non-atomic congestion games.

2.6 Load Balancing with Unrelated Machines

A class of optimization problems, whose game-theoretic interpretation is strictly connected to congestion games and load balancing games, is that of *load balancing problems with unrelated machines*.[5]

Definition 2.13 (Awerbuch et al. [11]). A *load balancing instance with unrelated machines* is a tuple $\mathsf{LB} = (I, J, (w_{ij})_{i \in I, j \in J})$ such that I is a set of m machines and $w_{ij} \in [0, \infty]$ is the *weight* of *job* j when assigned to *machine* i.

By strengthening Definition 2.13, we get several interesting classes of load balancing instances:

- if each weight w_{ij} is defined as w_j/s_i, where $w_j \geq 0$ is the weight of job j (not depending on the selected machine) and $s_i > 0$ is the *speed* of machine i, we get a *load balancing instance with related machines*;

[5] For the general model of load balancing considered in this section, we use a partially different notation with respect to that used for (load balancing) congestion games, and we adopt a notation that is commonly used in the field of machine scheduling. Anyway, in the remainder of the book, it will be clear which model of load balancing (and related notation) we are considering; furthermore, this general model of load balancing will be treated in the "Related Work" section of this Chapter and in Section 4.4, only.

- if w_{ij} can be equal either to ∞ or to w_j/s_i, we get a *linear load balancing instance with related machines and restricted assignments*;
- if w_{ij} can be equal either to ∞ or to w_j, we get a *load balancing instance with identical machines*.

A *feasible assignment* is a n-uple $\boldsymbol{\sigma} = (\sigma_1, \sigma_2, \ldots, \sigma_n)$ such that $\boldsymbol{\sigma}_j \in I$ (i.e., $i := \boldsymbol{\sigma}_j$ is the machine which job $j \in J$ has been assigned under $\boldsymbol{\sigma}$), and $w_{\sigma_j, j} \neq \infty$ for any $j \in J$ (i.e., a job cannot be assigned to a machine for which its weight is ∞). Given $i \in I$,

$$w_i(\boldsymbol{\sigma}) = \sum_{j \in J_i(\boldsymbol{\sigma})} w_{ij}$$

is the *load* of machine i under assignment $\boldsymbol{\sigma}$, where $J_i(\boldsymbol{\sigma}) := \{j \in J : \sigma_j = i\}$ denotes the set of jobs assigned to machine i in $\boldsymbol{\sigma}$. In the context of machine scheduling, the load of a machine i in $\boldsymbol{\sigma}$ models the time needed to complete all the jobs assigned to i (in $\boldsymbol{\sigma}$).

The *(offline) load balancing problem* consists in finding a feasible assignment $\boldsymbol{\sigma}$ minimizing a *total load function* $\mathsf{SF}(\boldsymbol{\sigma})$ which is non-decreasing in each load $w_i(\boldsymbol{\sigma})$. In the *online linear load balancing problem*, each job j arrives online, and should be irrevocably assigned to some machine i with the aim of minimizing as much as possible SF and without knowing the characteristics of the jobs related to future arrivals.

Commonly studied total load functions are:

- the *makespan* (or *L_∞-norm of the total load*), defined as $\mathsf{TL}_\infty(\boldsymbol{\sigma}) = \max_{i \in I} w_i(\boldsymbol{\sigma})$;
- the *L_p norm of the total load*, defined as $\mathsf{TL}_p(\boldsymbol{\sigma}) = \sqrt[p]{\sum_{i \in I} w_i^p(\boldsymbol{\sigma})}$ for some $p \geq 1$;
- as a generalization of the L_p norm of the total load, we have the *f-weighted total load*, defined as $\mathsf{WTL}_f(\boldsymbol{\sigma}) = \sum_{i \in I} w_i(\boldsymbol{\sigma}) f(w_i(\boldsymbol{\sigma}))$ for a given increasing function $f : \mathbb{R}_{\geq 0} \to \mathbb{R}_{\geq 0}$ with $f(0) = 0$; we observe that, if f is a monomial of degree p, the f-weighted total load coincides with the $(p-1)$th power of the L_{p-1} norm.

A classic online algorithm used for this problem is the *greedy algorithm* (Awerbuch et al. [11], Graham [98]): each time a job arrives, this algorithm assigns it to the machine causing the lowest increase in function SF.

By considering a game-theoretic version of the previous optimization problems, we obtain *load balancing games with unrelated machines*: each job j is associated to a selfish player who can choose a machine i which j should be assigned to, with the aim of minimizing the load $w_i(\boldsymbol{\sigma})$ of the machine $i = \sigma_j$ she chooses. As in congestion games, such a selfish behavior leads to *pure Nash equilibria*, i.e. stable outcomes in which no player can reduce her completion time up to a factor $1+\varepsilon$ by unilaterally deviating in favor of another machine.

Observe that load balancing games with related machines can be modeled as symmetric weighted load balancing games (Section 2.2), and load balancing games with related machines and restricted assignments can be modeled as more general weighted load balancing games. Analogously to Section 2.2, one can define for these games the solution concepts of ϵ-approximate pure Nash equilibria and ϵ-approximate one-round walks, and the efficiency metrics of ϵ-approximate prices of anarchy and stability, and the competitive ratio of ϵ-approximate one round walks. Observe that the greedy algorithm defined above can be obtained by simulating the cooperative one-round walk.

2.7 Related Work

In this section, we give a brief overview of the state-of-the-art in congestion games and their performance analysis, as it is strongly related to the results presented in this book. Furthermore, we also provide some relevant references in the fields of offline and online optimization for load balancing and other resource selection problems, since part of our findings can be applied to these contexts.

Price of Anarchy and Price of Stability

Atomic Games

The price of anarchy in (atomic) congestion games was first considered by Awerbuch et al. [12] and Christodoulou and Koutsoupias [59], who independently show that the price of anarchy is $5/2$ in unweighted games with affine latency functions. Christodoulou and Koutsoupias [59] also prove that better bounds are not possible in both symmetric unweighted games and unweighted network games; these results were improved by Correa et al. [76] who show that the price of anarchy stays the same even in symmetric unweighted network games. Furthermore, Awerbuch et al. [12] additionally show that the price of anarchy of weighted congestion games with affine latency functions is $(3+\sqrt{5})/2$.

Caragiannis et al. [54] show that the previous bounds are tight also for load balancing games. For the special case of load balancing games on identical resources, the works of Suri et al. [170] and Caragiannis et al. [54] imply that the price of anarchy is 2.012067 for unweighted games and at least $5/2$ for weighted ones. Paccagnan et al. [140] and Ravindran Vijayalakshmi and Skopalik [152] show how to obtain tight bounds for the price of anarchy of unweighted games with identical resource having polynomial latencies, and they provide some numerical values for the first degrees.

In Lücking et al. [120], it is proved that, for symmetric load balancing games, the price of anarchy drops to $4/3$ in unweighted games, and to $9/8$ in weighted games with identical resources. This last result has been generalised by Gairing et al. [93] who provide tight bounds on the price of anarchy of symmetric weighted load balancing games with identical resources and monomial latency functions. For symmetric unweighted k-uniform matroid congestion games with affine latency functions, de Jong et al. [81] prove that the price of anarchy is at most $28/13$ and at least 1.343 for a sufficiently large value of k (for $k = 5$, it is roughly 1.3428).

Tight bounds of $\Theta\left(\frac{p}{\log(p)}\right)^{p+1}$ on the price of anarchy of either weighted and unweighted congestion games with polynomial latency functions have been given by Aland et al. [7].[6] Under fairly general latency functions, Fotakis [88] shows that the price of anarchy of unweighted symmetric load balancing games coincides with that of non-atomic congestion games, while Bhawalkar et al. [20] prove that assuming symmetric strategies does not lead to improved bounds in unweighted games and give exact bounds for the case of weighted players. It also shown that, for the case of weighted players, no improvements are possible even in symmetric load balancing games with monomial and polynomial latency functions. Gairing and Schoppmann [92] and Gairing et al. [93, 94] provide upper and lower bounds on the pure and mixed price of anarchy for several classes of load balancing games with polynomial latency functions. Finally, Christodoulou et al. [61] (resp. Bilò [22]) provide tight bounds (resp. almost tight upper bounds) on the approximate price of anarchy of unweighted (resp. weighted) congestion games under affine latency functions. Giannakopoulos et al. [95] have recently obtained tight bounds on the approximate price of anarchy of congestion games with polynomial latency functions, holding even for load balancing games; prior to this work, analogue results were obtained by Bilò and Vinci [24].

Roughgarden [158] characterizes the price of anarchy for unweighted congestion games having general non-decreasing latency functions and proves that the price of anarchy of congestion games under pure Nash equilibria is as high as that under mixed/correlated/coarse correlated Nash equilibria. This last result is extended to weighted games by Bhawalkar et al. [20] and to approximate equilibria and even non-decreasing latency functions by Bilò [42]. Roughgarden [157] investigates the quality of Nash equilibria in congestion games with incomplete informations.

Relatively to the price of stability, Christodoulou and Koutsoupias [60] give a lower bound of $1 + \frac{1}{\sqrt{3}}$ for affine unweighted congestion games, and Caragiannis et al. [54] prove that this bound is tight. Christodoulou and Gairing [67] prove a tight bound of $\Theta(p)$ on the price of stability for polynomial latency functions of maximum degree p, which is, thus, asymptotically better than the corresponding price of anarchy. In contrast, Christodoulou et al. [65] show that, for weighted games and polynomial latencies, the price of stability is asymptotically close to the price of anarchy. Christodoulou et al. [61] and Bilò [22] give bounds for the ε-approximate price of stability of affine unweighted and weighted congestion games respectively. Anshelevich et al. [8] show that the price of stability of load balancing games under fairly general latency functions, coincides with the price of anarchy of non-atomic congestion games having the same latency functions. The price of stability has been widely investigated in congestion games with non-increasing latency functions (e.g.,

[6] For the sake of conciseness, most of the bounds related to polynomial latency functions are represented via their asymptotic growth (in the maximum degree of the considered polynomials). To see the exact values, refer to the related bibliography.

cost-sharing games; see Anshelevich et al. [8], Bilò et al. [32, 38], Fiat et al. [86], Li [118]).

See Table 2.1 for a compact representation of some bounds on the price of anarchy and price of stability discussed in this subsection.

Non-atomic Games

Roughgarden and Tardos [160] prove that the price of anarchy of non-atomic congestion games with affine latency functions is $4/3$. Roughgarden [155] extends this result to polynomial latency functions of degree p, and shows that the price of anarchy, in this case, is $\Theta\left(\frac{p}{\log(p)}\right)$. Correa et al. [79], Roughgarden [155] and Roughgarden and Tardos [161] characterize the price of anarchy for very general latency functions (by providing closed-form bounds), and show that this is tight for simple Pigou's networks (i.e., symmetric games with two resources). Christodoulou et al. [61] give tight bounds on the ε-approximate price of anarchy and stability for congestion games with polynomial latency functions.

See the table illustrated in Figure 2.1 to compare the price of anarchy and price of stability of atomic and non-atomic congestion games.

	U/CG	U/LB	U/SLB	U/Id/LB	N/CG
PoA(1)	$\frac{5}{2}$ [12, 59]	$\frac{5}{2}$ [54]	$\frac{4}{3}$[120]	2.012 [54, 170]	$\frac{4}{3}$ [159, 161]
PoA(p)	$\Theta\left(\frac{p}{\log(p)}\right)^{p+1}$ [7]	$\Theta\left(\frac{p}{\log(p)}\right)^{p+1}$ [92]	$\Theta\left(\frac{p}{\log(p)}\right)$ [88]	[140, 152]	$\Theta\left(\frac{p}{\log(p)}\right)$ [155]
PoS(1)	$1+\frac{1}{\sqrt{3}}$ [54, 60]	$\frac{4}{3}$ [8]	$\frac{4}{3}$ [8]	X	$\frac{4}{3}$ [159, 161]
PoS(p)	$\Theta(p)$ [67]	$\Theta\left(\frac{p}{\log(p)}\right)$ [8]	$\Theta\left(\frac{p}{\log(p)}\right)$ [8]	X	$\Theta\left(\frac{p}{\log(p)}\right)$ [155]

	W/CG	W/LB	W/SLB	W/Id/LB	W/Id/SLB
PoA(1)	$\frac{3+\sqrt{5}}{2}$ [12]	$\frac{3+\sqrt{5}}{2}$ [54]	$\frac{3+\sqrt{5}}{2}$ [20]	$\frac{5}{2}^{(low)}$ [54]	$\frac{9}{8}$ [120]
PoA(p)	$\Theta\left(\frac{p}{\log(p)}\right)^{p+1}$ [7]	$\Theta\left(\frac{p}{\log(p)}\right)^{p+1}$ [20]	$\Theta\left(\frac{p}{\log(p)}\right)$ [20]	X	$\Theta\left(\frac{2^p}{p}\right)$ [93]
PoS(1)	$2^{(upp)}$ [22]	X	X	X	X
PoS(p)	$\Theta\left(\frac{p}{\log(p)}\right)^{p+1}$ [22]	X	X	X	X

Fig. 2.1 Bounds on the price of anarchy and the price of stability for atomic and non-atomic congestion games with affine and polynomial latency functions (for $\varepsilon = 0$). PoA(1) and PoA(p) refer to the price of anarchy of affine and polynomial latencies (of maximum degree p), respectively; PoS(1) and PoS(p) are defined accordingly for the price of stability. CG refers to general congestion games, LB refers to load balancing games, SLB refers to symmetric load balancing games, Id refers to games with identical resources, W refers to weighted atomic games, U refers to unweighted atomic games, and N refers to non-atomic games. We put apex "(upp)" (resp. "(low)") when the given bound may not be tight, and it is an upper (resp. lower) bound only. We put "X" to denote some cases which have not been studied or solved yet (to the best of our knowledge).

Approximate Social Optima and pure Nash Equilibria via the Price of Anarchy

The work on the quantification of the price of anarchy for certain classes of congestion games has also influenced the design of efficient algorithms for computing either approximate social optima or approximate pure Nash equilibria. Awerbuch et al. [13] resort to certain best-response dynamics applied to certain potential games (including polynomial unweighted and linear weighted congestion games), with the aim of computing some ε-approximate variants of pure Nash equilibria (where $\varepsilon > 0$ can be arbitrarily chosen); by using the definition of the price of anarchy, they show that such equilibria guarantee a social welfare of at most $\beta + \varepsilon$ times the social optimum, where β is the price of anarchy of the input game.

By exploiting similar best-response dynamics, Caragiannis et al. [53] compute $(\alpha + \varepsilon)$-approximate equilibria in affine and polynomial congestion games, where α is the price of anarchy of a game connected to the input one; in particular, they provide polynomial-time algorithms to compute $(2 + \varepsilon)$-approximate (resp. $p^{O(p)}$-approximate) equilibria for affine (resp. polynomial) unweighted congestion games. By using a similar machinery, Caragiannis et al. [55] provide similar approximation guarantees for weighted games, and Ravindran Vijayalakshmi and Skopalik [152] provide better approximation factors for unweighted ones. Other results on existence and computation of approximate equilibria have been also achieved by Caragiannis and Fanelli [51], Christodoulou et al. [66], Giannakopoulos et al. [95], Hansknecht et al. [99].

Atomic Games with Splittable Flow

An interesting variant of congestion games is that of *atomic congestion games with splittable flow*, in which each player has an amount of flow she controls which can be split among her strategies. As the flow is splittable, this class of games exhibits several similarities with its non-atomic counterpart. Cominetti et al. [74], Harks [100] and Roughgarden and Schoppmann [162] have studied the price of anarchy of atomic games with splittable flow; in particular, Roughgarden and Schoppmann [162] provide a tight bound for general latency functions which equals $\Theta(p)^{\frac{p+1}{2}}$ for the particular case of polynomial latencies of maximum degree p.

One-round Walk Starting From the Empty State

The competitive ratio of exact one-round walks generated by cooperative players in weighted load balancing games with polynomial latency functions has been implicitly considered by Awerbuch et al. [11], who analyzed the performance of the online greedy algorithm under the L_p-norm. Their results directly imply an upper bound of $3 + 2\sqrt{2}$ for the special case of affine functions, and an upper bound of $\Theta(p)^p$ for general polynomial latencies. For unweighted players, the result for affine functions has been improved to $17/3$ by Suri et al. [170], who also show that, for identical resources, the upper bound drops to $2+\sqrt{5}$ in spite of a lower bound of 3.0833. Finally, Caragiannis et al. [54] show matching lower bounds of $3+2\sqrt{2}$ and $17/3$ for, respectively, weighted and unweighted players with affine latencies. For weighted games with polynomial latency functions, a tight bound of $\Theta(p)^p$ is directly implied by Caragiannis [49]; the lower bounds, in particular, hold even for identical resources, thus improving previous results from Awerbuch et al. [11]. Caragiannis et al. [54] also show that, for unweighted players and identical resources, the competitive ratio lies between 4 and $\frac{2}{3}\sqrt{21}+1$ if the latency functions are affine.

For the case of selfish players and still under affine latency functions, Bilò et al. [31] and Christodoulou et al. [63] show that the competitive ratio is $2+\sqrt{5}$ for unweighted congestion games, while, for weighted players, Christodoulou et al. [63] give an upper bound of $4+2\sqrt{3}$. For the more general case of polynomial latency functions, Bilò and Vinci [24] and Klimm et al. [110] determine explicit and good upper bounds on the competitive ratio in unweighted and weighted congestion games.

Offline and Online Optimization for Load Balancing with Unrelated Machines

In the context of offline optimization, Lenstra et al. [116] and Shmoys and Tardos [167] design 2-approximation algorithms for makespan minimization in load balancing with unrelated machines; furthermore, Lenstra et al. [116] show that the problem is not approximable within a ratio better than $3/2$. In the case of related machines, Hochbaum and Shmoys [104] design a PTAS.

For online minimization, Bartal et al. [16] design an algorithm providing a competitive ratio slightly better than 2 (we observe that 2 is the competitive ratio resulting from the classical greedy algorithm of Graham [98]). Relatively to the online L_2-norm minimization, Awerbuch et al. [11] prove that the greedy algorithm has a competitive ratio of at most $1+\sqrt{2}$, and Caragiannis et al. [54] prove that this ratio cannot be improved.

For the general online L_p-norm minimization, Awerbuch et al. [11] provide upper bounds growing as $\Theta(p)$ for the competitive ratio of the greedy algorithm, and Caragiannis [48] gives tight bounds of $\frac{1}{2^{1/p}-1} \sim \Theta(p)$ for the competitive ratio of the greedy algorithm and also prove that it cannot be improved by any online algorithm, even for identical machines. Anyway, in the context of offline optimization, the approximation guarantee has been improved by Makarychev and Sviridenko [121], who provide a $\sqrt[p]{B_p+\varepsilon}$-approximation algorithm (for any $\varepsilon>0$), where B_p denotes the pth Bell number.

Chapter 3
Bounding the Inefficiency via the Primal-Dual Method

The main tool used to show all the results in this book is the *primal-dual method*. Introduced by Bilò [22], this framework is based on the construction and analysis of a primal-dual pair of linear programs, where the dual is used to provide tight (or almost tight) upper bounds on the performance guarantee of a certain efficiency metric, and the primal is used to construct tight (or almost tight) lower bounds.

In this chapter, we present the general scheme of the primal-dual method and then show its effectiveness by doing some comparisons with respect to other existing techniques (e.g., the smoothness framework of Roughgarden [158]). We also provide a warm-up example illustrating how to apply our machinery to a simple introductory game. Then, in the subsequent chapters of the book, we shall give further and more sophisticated applications; for instance, we shall apply it to congestion games with very general latency functions (e.g., those defined in Section 2.2), or to other interesting variants of congestion games.

For further details on linear programming duality, the reader can refer to the appendix (Section B.1) or to the textbooks by Matouek and Gärtner [124] and Boyd and Vandenberghe [43].

3.1 The Primal-Dual Method

In this section, we describe the primal-dual method by applying it to a generic class of congestion games, in which the underlying solution concepts, social functions and efficiency metrics are not specified and given as input. Assume we want to establish an upper bound on the worst-case performance guarantee of a solution concept SC (e.g., pure Nash equilibria) with respect to a social function SF (e.g., total latency) and an efficiency metric EM (e.g., price of anarchy) in a class of games $\mathcal{G}$.

V. Bilò, C. Vinci, *Coping with Selfishness in Congestion Games*, Monographs in Theoretical Computer Science. An EATCS Series, https://doi.org/10.1007/978-3-031-30261-9_3

Primal Formulation

By fixing an arbitrary game $\mathsf{CG} \in \mathcal{G}$, a social optimum $\boldsymbol{\sigma}^*$ and a strategy profile $\boldsymbol{\sigma} \in \mathsf{SC}(\mathsf{CG})$, where $\mathsf{SC}(\mathsf{CG})$ denotes the set of strategy profiles of CG implementing the considered solution concept SC, the method requires to construct a linear program $\mathsf{LP}_{\boldsymbol{\alpha}}(\mathsf{SF}, \boldsymbol{\sigma}, \boldsymbol{\sigma}^*)$ in a vector of variables $\boldsymbol{\alpha} := (\alpha_e)_{e \in E}$ defined as

$$\max \quad \mathsf{SF}_{\boldsymbol{\alpha}}(\boldsymbol{\sigma}) \tag{3.1}$$

$$s.t. \quad constraints_{\boldsymbol{\alpha}}(\boldsymbol{\sigma}, \boldsymbol{\sigma}^*) \tag{3.2}$$

$$\mathsf{SF}_{\boldsymbol{\alpha}}(\boldsymbol{\sigma}^*) = 1 \tag{3.3}$$

$$\alpha_e \geq 0 \quad \forall e \in E$$

and satisfying the following properties:

- $\mathsf{SF}_{\boldsymbol{\alpha}}(\boldsymbol{\sigma})$ and $\mathsf{SF}_{\boldsymbol{\alpha}}(\boldsymbol{\sigma}^*)$ are linear functions in the variables α_es, that coincide with $\mathsf{SF}(\boldsymbol{\sigma})$ and $\mathsf{SF}(\boldsymbol{\sigma}^*)$ when all α_es are equal to 1 (i.e., $\mathsf{SF}_{\mathbf{1}}(\boldsymbol{\sigma}) = \mathsf{SF}(\boldsymbol{\sigma})$ and $\mathsf{SF}_{\mathbf{1}}(\boldsymbol{\sigma}^*) = \mathsf{SF}(\boldsymbol{\sigma}^*)$). For instance, $\mathsf{SF}_{\boldsymbol{\alpha}}$ can be obtained from SF by multiplying the latency of each resource e by α_e;
- $constraints_{\boldsymbol{\alpha}}(\boldsymbol{\sigma}, \boldsymbol{\sigma}^*)$ is a set of $s \geq 1$ linear constraints that, for $\boldsymbol{\alpha} = \mathbf{1}$, capture the properties possessed by the target solution concept (or a part of them). For instance, when considering ε-approximate equilibria, we can use the argument that the cost of player i in $\boldsymbol{\sigma}$ is no more than $(1+\varepsilon)$ times the cost that i can suffer when deviating to strategy σ_i^*, and this leads to n inequalities (one for each player); then, we can multiply the latency of each $e \in E$ by α_e (in each inequality), thus providing n linear inequalities/constraints;
- (3.3) is a normalization constraint imposing value 1 to the social optimum. This shall imply that the maximum value of the objective function (3.1) is equal to the highest ratio $\mathsf{SF}_{\boldsymbol{\alpha}}(\boldsymbol{\sigma})/\mathsf{SF}_{\boldsymbol{\alpha}}(\boldsymbol{\sigma}^*) = \mathsf{SF}_{\boldsymbol{\alpha}}(\boldsymbol{\sigma})$.

We observe that, whenever $\mathsf{SF}(\boldsymbol{\sigma}^*) \neq 0$ (if this is not the case, one can generally show that the value of the considered efficiency metric is unbounded), the normalization considered in (3.3) is well-defined and does not affect the value of the ratio $\mathsf{SF}(\boldsymbol{\sigma})/\mathsf{SF}_{\boldsymbol{\alpha}}(\boldsymbol{\sigma}^*)$ (as both the numerator and the denominator are linear in the α_es).

Now, by the arbitrariness of $\boldsymbol{\sigma}$, $\boldsymbol{\sigma}^*$ and CG, we have that the optimal value $\mathsf{SF}_{\boldsymbol{\alpha}}(\boldsymbol{\sigma}) = \mathsf{SF}_{\boldsymbol{\alpha}}(\boldsymbol{\sigma})/\mathsf{SF}_{\boldsymbol{\alpha}}(\boldsymbol{\sigma}^*)$ is an upper bound on the worst-case performance of the considered class of congestion games, with respect to the considered efficiency metric EM.

A specific instantiation of the above general linear program has been provided in Example 3.1, where the goal is to determine the price of anarchy of affine unweighted congestion games.

Dual Formulation

Now, to provide a tight (or almost tight) upper bound, the method requires to derive and analyze the dual of the above linear program which has the following representation:

$$\begin{aligned} \min \quad & \gamma \\ s.t. \quad & dualconstraints_e((x_t)_{t\in[s]}, \gamma), \quad \forall e \in E \\ & x_t \geq 0 \quad \forall t \in [s], \end{aligned}$$

where each dual variable x_t is associated to a constraint in (3.2), γ is the dual variable associated to the normalization constraint (3.3), and $dualconstraints_e((x_t)_{t\in[s]}, \gamma)$ is the dual constraint related to the primal variable α_e. By providing a feasible solution to the dual formulation, an upper bound on the worst-case performance guarantee of the target solution concept is obtained. Indeed, by the Weak Duality Theorem[1], the value of the objective function of the dual program is at least the maximum value of the objective function of the primal.

3.2 The Effectiveness of the Primal-Dual Method

An advantage of using the primal-dual method is that, for several efficiency metrics, and for several variants of congestion games, it is generally easy to determine a primal program (of the type described above) whose optimal solution provides an upper bound on the considered efficiency metric. Despite it is usually difficult to obtain this optimal solution, the dual formulation, instead, immediately provides some inequalities $dualconstraints_e((x_t)_{t\in[s]}, \gamma)$ that can be translated into compact and quasi-explicit formulas yielding an upper bound γ on the worst-case performance of the considered class of congestion games that is possibly tight. Moreover, the proof of correctness of these formulas is automatically given by the correct definition of the initial LP formulation and by the Weak Duality Theorem.

Smoothness Framework and the Primal-Dual Method

The quasi-explicit formulas derived by inspecting the dual are similar (and in some cases equivalent) to those exploited within the *smoothness framework*[2]: an effective machinery introduced and formalized by Roughgarden [158] with the aim of unifying and generalizing several existing approaches adopted to quantify the performance degradation under selfish behavior. In particular, the smoothness-framework can be used to derive tight (or almost tight) bounds on the efficiency of several classes of games (including congestion games), and it is based on an inequality linking together the social optimum and the sum of the players' costs in an equilibrium, thus requiring the use of two variables. However, for certain settings, additional structural properties of the game need to be embedded in the model. This requires more sophisticated constraints involving a larger number of variables.

[1] For further details on the Weak Duality Theorem, see Matouek and Gärtner [124] or Section B.1 of the appendix.

[2] A formal description of the smoothness framework is deferred to Section B.2 of the appendix.

The primal-dual method handles these twists more easily, as it suffices writing in the primal program all the additional constraints that need to be satisfied by the model, and then the related dual program automatically provides the quasi-explicit formulas that characterize the value of the considered efficiency metric[3].

From Tight Upper Bounds to Tight Lower Bounds via Strong Duality

Another advantage of the primal-dual method is that, the more $constraints_{\boldsymbol{\alpha}}(\boldsymbol{\sigma}, \boldsymbol{\sigma}^*)$ provide an accurate characterization of the properties of the target solution concept, the more the achieved upper bound will be significant (and possibly tight). Thus, by considering an optimal solution of the dual program, and by resorting to the Strong Duality Theorem and the complementary slackness conditions[4], one can find an optimal solution to the primal program that can be used as an initial hint to construct tight lower bounds on the considered efficiency metric.

An Introductory Example

To highlight the key-features of the primal-dual method and its advantages, we illustrate its application to the characterization of the exact price of anarchy of unweighted congestion games with affine latency functions, thus reobtaining the known result of Christodoulou and Koutsoupias [59] and Awerbuch et al. [12].

Example 3.1 (PoA of Affine Unweighted Congestion Games [12, 59]). Given an arbitrary affine unweighted atomic congestion game CG, a pure Nash equilibrium $\boldsymbol{\sigma}$ and an optimal strategy profile $\boldsymbol{\sigma}^*$, the following linear program in variables α_es and β_es is clearly an upper bound on the price of anarchy (with respect to the total latency function):

$$\max \quad \mathsf{TL}(\boldsymbol{\sigma}) = \sum_{e\in E}(\alpha_e k_e^2 + \beta_e k_e) \tag{3.4}$$

$$s.t. \quad \sum_{e\in\sigma_i}(\alpha_e k_e + \beta_e) \le \sum_{e\in\sigma_i^*}(\alpha_e(k_e+1)+\beta_e) \quad \forall i \in N \tag{3.5}$$

$$\mathsf{TL}(\boldsymbol{\sigma}^*) = \sum_{e\in E}(\alpha_e o_e^2 + \beta_e o_e) = 1 \tag{3.6}$$

$$\alpha_e, \beta_e \ge 0 \quad \forall e \in E.$$

Indeed, variables α_es and β_es are such that the latency function ℓ_e of each resource e is of the form $\ell_e(x) := \alpha_e \cdot x + \beta_e$, i.e., it is an arbitrary affine function. k_e and o_e are the congestions of resource e in $\boldsymbol{\sigma}$ and $\boldsymbol{\sigma}^*$, respectively. Observe that k_e^2, o_e^2, k_e and o_e have to be considered as fixed parameters. Constraints (3.5) impose necessary conditions for $\boldsymbol{\sigma}$ to be a pure Nash equilibrium, constraint (3.6) imposes without

[3] Several examples in which the application of the primal-dual method generates dual programs with more than two variables are given in Chapters 5 (Theorem 5.5) and 6 (Theorem 6.1).

[4] For further details on the Strong Duality Theorem and the complementary slackness conditions, see Matouek and Gärtner [124] or Section B.1 of the appendix

loss of generality that the optimal social function $\mathsf{TL}(\boldsymbol{\sigma}^*)$ is equal to 1, and the objective function (3.4) is equal to the social function $\mathsf{TL}(\boldsymbol{\sigma})$ of $\boldsymbol{\sigma}$. Because of the normalization constraint (3.6), the objective function is also equal to the ratio $\mathsf{TL}(\boldsymbol{\sigma})/\mathsf{TL}(\boldsymbol{\sigma}^*)$, and the maximum value of the considered linear program over all possible strategy profiles $\boldsymbol{\sigma}$ and $\boldsymbol{\sigma}^*$ and congestion games CGs is clearly an upper bound on $\mathsf{TL}(\boldsymbol{\sigma})/\mathsf{TL}(\boldsymbol{\sigma}^*)$, i.e., an upper bound on the price of anarchy of the class of affine unweighted atomic congestion games.

By summing all constraints in (3.2) we obtain a relaxed linear program in which all constraints in (3.5) are replaced with constraint

$$\sum_{e\in E}(\alpha_e k_e^2+\beta_e k_e)\leq \sum_{e\in E}(\alpha_e o_e(k_e+1)+\beta_e o_e). \tag{3.7}$$

The dual of this new linear program is defined in two variables x,γ and has the following form:

$$\begin{aligned} \min\quad & \gamma \\ s.t.\quad & xk_e^2-xo_e(k_e+1)+\gamma o_e^2\geq k_e^2, \quad \forall e\in E \\ & xk_e-xo_e+\gamma o_e\geq k_e, \quad \forall e\in E \\ & x\geq 0. \end{aligned} \tag{3.8}$$

Solution $(x,\gamma)=\left(\frac{3}{2},\frac{5}{2}\right)$ is feasible, thus $\frac{5}{2}$ is an upper bound on the price of anarchy of affine congestion games. It is also optimal, as it can be obtained by imposing that (3.8) is tight (i.e., satisfied at equality) for $(k_e,o_e)=(1,1)$ and $(k_e,o_e)=(2,1)$. Furthermore, in the dual program, we can avoid all non-tight constraints, and we can simply use the (tight) dual constraints of type (3.8) related to $(k_e,o_e)=(1,1)$ and $(k_e,o_e)=(2,1)$ respectively, without affecting the resulting optimal solution. Thus, by applying the Strong Duality Theorem to the new dual program, we have that the upper bound $\frac{5}{2}$ can be also obtained as optimal solution of the relaxed primal program. Furthermore, since $x>0$, the primal constraint (3.7) is tight (by the complementary slackness conditions). We conclude that the maximum value of the objective function in the following linear program is $\frac{5}{2}$:

$$\begin{aligned} \max\quad & \alpha_1k_1^2+\alpha_2k_2^2 && (3.9)\\ s.t.\quad & \alpha_1k_1^2+\alpha_2k_2^2=\alpha_1o_1(k_1+1)+\alpha_2o_2(k_2+1) && (3.10)\\ & \alpha_1o_1^2+\alpha_2o_2^2=1 && (3.11)\\ & \alpha_1,\alpha_2\geq 0, \end{aligned}$$

where $(k_1,o_1)=(1,1)$ and $(k_2,o_2)=(2,1)$. The optimal solution (leading to the optimal value $\frac{5}{2}$) is given by $(\alpha_1,\alpha_2)=\left(\frac{1}{2},\frac{1}{2}\right)$, and it is univocally determined by equalities (3.10) and (3.11). Now, starting form the previous linear program and its optimal solution, we construct a congestion game CG satisfying the following *tight conditions*:

- each player $i\in N$ has two strategies σ_i,σ_i^* only;

- the cost of each player i in strategy profile $\boldsymbol{\sigma}$ is equal to the objective function (3.9), and then, it is also equal to the left-hand part of constraint (3.10);
- the cost of player i, when she unilaterally deviates to strategy σ_i^* from strategy profile $\boldsymbol{\sigma}$, is equal to the right-hand part of constraint (3.10), thus $\boldsymbol{\sigma}$ is a pure Nash equilibrium;
- the cost of each player i in strategy profile $\boldsymbol{\sigma}^*$ is equal to the left-hand part of constraint (3.11).

If we construct a congestion game CG satisying the previous tight conditions, the price of anarchy of CG will be equal to $\frac{5}{2}$, thus it will be a matching lower bound. A very natural choice for such lower bound is a particular case of the class of lower bounds provided by Aland et al. [7], who consider the more general case of unweighted games with polynomial latencies. For affine latencies, the above lower bound has the following structure:

- CG has a set of players $N := [3]$ and a set of resources $E := E^1 \cup E^2$, with $E^1 := \{e_1^1, e_2^1, e_3^1\}$ and $E^2 := \{e_1^2, e_2^2, e_3^2\}$;
- all the resources e of E^1 (resp. E^2) have latency function $\ell_1(x) := \alpha_1 x$ (resp. $\ell_2(x) := \alpha_2 x$), where $(\alpha_1, \alpha_2) = \left(\frac{1}{2}, \frac{1}{2}\right)$ is the optimal solution of the previous linear program (thus, in this particular case, it suffices that all the resources have the same linear latency function);
- each player i has strategies $\sigma_i := \{e_{i+1}^1, e_{i+1}^2, e_{i+2}^2\}$ and $\sigma_i^* := \{e_i^1, e_i^2\}$, where the resources' indices are supposed to be cyclical (i.e. $e_{i+3}^j := e_i^j$).

We observe that, when a player i selects strategy σ_i (resp. σ_i^*), she selects one resource from E_1 and two resources from E_2 (resp. one resource from E_1 and one resource from E_2); furthermore, pair $(k_1, o_1) = (1,1)$ (resp. $(k_2, o_2) = (2,1)$) is such that k_1 and o_1 (resp. k_2 and o_2) are the congestions of each resource $e \in E_1$ (resp. $e \in E_2$) under strategy profiles $\boldsymbol{\sigma}$ and $\boldsymbol{\sigma}^*$, respectively.

By simple computations, one can easily verify that CG satisfies all the above tight conditions, sufficient to have a matching lower bound of $\frac{5}{2}$ on the price of anarchy. Note that the congestion game CG considered in this example, despite already used in [7], has been here "reinterpreted" in terms of the hints given by the primal-dual method.

3.3 Concluding Remarks and Related Work

The primal-dual method has been introduced and formalized in the seminal paper of Bilò [22], and then further applied and explored in several papers co-authored by the authors of this book [18, 21, 23, 24, 26–29, 34–37, 42, 174, 175].

Other techniques for bounding the price of anarchy in non-cooperative games based on primal-dual formulations have been proposed by Kulkarni and Mirrokni [114], Nadav and Roughgarden [132] and Thang [172]. Similarly to the primal-dual method, the technique proposed by Nadav and Roughgarden [132] operates by explicitly formulating the problem of maximizing the price of anarchy of a class of games. However, despite using the same formulation, the two approaches differ in the choice of the variables: while the primal-dual method relies on suitable multipliers for the resource cost functions, Nadav and Roughgarden [132] use the probability distributions defining the outcomes occurring in the formulation. The methods in Kulkarni and Mirrokni [114] and Thang [172], instead, build on a formulation for the problem of optimizing the social function, and then implement the equilibria conditions within the choice of the dual variables.

Further generalizations or variants of the above primal-dual techniques have been also considered in the context of algorithmic design, with the aim of improving the Price of Anarchy in non-cooperative games (see, for instance, the works of Paccagnan et al. [140], Ravindran Vijayalakshmi and Skopalik [152]). With this respect, Paccagnan et al. [139] and Paccagnan and Marden [138] propose a general primal-dual framework to design the players utility/cost functions so as to optimize the Price of Anarchy, and such functions are defined as solutions of some tractable linear programs.

Part III
Analysis of Selfish and Greedy Outcomes in Congestion Games

Chapter 4
Congestion and Load Balancing Games with General Latency Functions

In this chapter, we analyze the performance of congestion games and load balancing games with very general latency functions.

In Section 4.1, we resort to the primal-dual method to provide explicit (or quasi-explicit) parametric formulas to compute upper bounds on the performance of weighted congestion games with general latency functions. The efficiency metrics we consider are the approximate price of anarchy and the approximate one-round walks involving both selfish and cooperative players.

Then, in Section 4.2, as the main result of the chapter, we show that the bounds provided in Section 4.1 are tight even for load-balancing games (under mild assumptions on the considered latency functions). Specifically, we show the following results:

- If $\mathcal{C}$ is closed under abscissa and ordinate scaling, the approximate price of anarchy does not improve when restricting to load balancing games (see Theorem 4.2). Furthermore, if $\mathcal{C}$ contains unbounded functions only (except for the eventual constant latency functions), the approximate price of anarchy does not improve even for symmetric load balancing games.
- Under latency functions drawn from $\mathcal{C}$, the competitive ratio of approximate one-round walks generated by selfish players does not improve when restricting to load balancing games (see Theorem 4.3).
- If all functions in $\mathcal{C}$ are semi-convex, the same limitation applies to the competitive ratio of approximate one-round walks generated by cooperative players (see Theorem 4.3).

The results presented in Section 4.2 are obtained by designing some general load balancing instances, which are parametrized by at most four numbers and two latency functions belonging to $\mathcal{C}$. We show that there exists a choice of the above parameters such that the performance of the considered load balancing instances matches that of general congestion games with latency functions in $\mathcal{C}$. Independently from our existential results, the load balancing instances we have introduced may be useful to get good lower bounds on the performance of load balancing games

V. Bilò, C. Vinci, *Coping with Selfishness in Congestion Games*, Monographs in Theoretical Computer Science. An EATCS Series, https://doi.org/10.1007/978-3-031-30261-9_4

with latency functions in $\mathcal{C}$, even if we are not able to quantify the exact value of the metric adopted to measure the performance (see Remark 4.4).

In Section 4.3, we apply the above general results to the specific case of polynomial latency functions, and we quantify the resulting approximate price of anarchy and the competitive ratio (see Table 4.2 for some numerical comparisons).

In Section 4.4, we apply the machinery of Section 4.1 to provide upper bounds on the efficiency of the f-weighted total load in load balancing games/problems with unrelated machines.

Finally, in Section 4.5, we analyze the subclass of unweighted games, and we obtain similar negative results as in the weighted case. In particular, we give the main intuitions on how to apply the previous approaches to the unweighted case, and we provide tight closed-form bounds on the efficiency of unweighted games with general latency functions. These bounds hold even for load balancing games if the underlying latencies are ordinate scaling. Again, we instantiate the obtained results to the case of polynomial latency functions (see Table 4.3 for some numerical comparisons).

4.1 Upper Bounds for General Congestion Games

In this section, we provide a theorem and a technical lemma, whose proofs are both based on the primal-dual method. Theorem 4.1 gives quasi-explicit formulas to compute upper bounds for the considered efficiency metrics, and then Lemma 4.1 provides some parameters that will be used in Section 4.2 to design tight load balancing lower bounds. More specifically, by following the primal-dual approach, we start with a primal program constructed from an arbitrary congestion game, and by exploiting its dual, we get the upper bounds of Theorem 4.1; as a complementary result, the parameters provided by Lemma 4.1 are derived by analyzing a simplified version of the initial primal program.[1]

We first give some preliminary definitions. Given $x \geq 1$, $k > 0$, $o > 0$, and a latency function f, let

$$\beta_{\mathsf{W}}(\mathsf{EM},k,o,f) := \begin{cases} -kf(k) + (1+\varepsilon)of(k+o) & \text{if } \mathsf{EM} = \mathsf{PoA}_{\varepsilon}, \\ -\int_0^k f(t)dt + (1+\varepsilon)of(k+o) & \text{if } \mathsf{EM} = \mathsf{CR}^s_{\varepsilon}, \\ -kf(k) + (1+\varepsilon)((k+o)f(k+o) - kf(k)) & \text{if } \mathsf{EM} = \mathsf{CR}^c_{\varepsilon}. \end{cases} \tag{4.1}$$

Given $\mathsf{EM} \in \{\mathsf{PoA}_{\varepsilon}, \mathsf{CR}^s_{\varepsilon}, \mathsf{CR}^c_{\varepsilon}\}$, let

$$\gamma_{\mathsf{W}}(\mathsf{EM},x,k,o,f) := \frac{kf(k) + x \cdot \beta_{\mathsf{W}}(\mathsf{EM},k,o,f)}{of(o)}; \tag{4.2}$$

[1] We point out that the upper bounds provided in Theorem 4.1 can be equivalently derived by resorting to the smoothness framework [158] (refer to Section B.2 in the appendix). Furthermore, similar results as in Lemma 4.1 have been also considered in [20, 158], with the aim of designing parametric load balancing instances whose performance match the upper bounds.

furthermore, given a class of weighted congestion games $\mathcal{G}$, let

$$\gamma_{\mathsf{W}}(\mathsf{EM},\mathcal{G}) := \inf_{x\geq 1} \sup_{f\in\mathcal{C}(\mathcal{G}),k,o>0} \gamma_{\mathsf{W}}(\mathsf{EM},x,k,o,f). \tag{4.3}$$

Theorem 4.1. *Let $\mathcal{G}$ be a class of weighted congestion games. For any efficiency metric $\mathsf{EM} \in \{\mathsf{PoA}_\varepsilon, \mathsf{CR}^s_\varepsilon\}$, we have that $\mathsf{EM}(\mathcal{G}) \leq \gamma_{\mathsf{W}}(\mathsf{EM},\mathcal{G})$. This fact holds for $\mathsf{EM} = \mathsf{CR}^c_\varepsilon$ if the latency functions of $\mathcal{C}(\mathcal{G})$ are semi-convex.*

Proof. We prove the theorem for $\mathsf{EM} = \mathsf{PoA}_\varepsilon$ by using the primal-dual approach, and we give a sketch for the other cases, since the proof is analogue. Let $\mathsf{CG} \in \mathcal{G}$, and let $\boldsymbol{\sigma}$ and $\boldsymbol{\sigma}^*$ be an ε-approximate pure Nash equilibrium and an optimal strategy profile, respectively; for brevity, we use k_e and o_e in place of $k_e(\boldsymbol{\sigma})$ and $k_e(\boldsymbol{\sigma}^*)$, respectively. We have that the maximum value of the following linear program in the variables α_es is an upper bound on $\mathsf{PoA}_\varepsilon(\mathsf{CG})$:

$$\begin{aligned} \mathsf{LP1}: \quad \max \quad & \overbrace{\sum_{e\in E} \alpha_e k_e \ell_e(k_e)}^{\mathsf{SUM}(\boldsymbol{\sigma})} \\ s.t. \quad & \sum_{e\in\sigma_i} \alpha_e \ell_e(k_e) \leq (1+\varepsilon) \sum_{e\in\sigma_i^*} \alpha_e \ell_e(k_e+w_i), \quad \forall i \in N && (4.4) \\ & \overbrace{\sum_{e\in E} \alpha_e o_e \ell_e(o_e)}^{\mathsf{SUM}(\boldsymbol{\sigma}^*)} = 1 && (4.5) \\ & \alpha_e \geq 0, \quad \forall e \in E. \end{aligned}$$

Indeed, by setting $\alpha_e = 1$ for any $e \in E$, we have that:

- the objective function is the social cost at the equilibrium;
- the constraints in (4.4) impose some relaxed ε-approximate pure Nash equilibrium conditions (ensuring that each player, at the equilibrium, does not get any benefit when deviating in favor of strategy σ_i^*);
- (4.5) is the normalized optimal social cost (normalization is possible since there is some $o_e > 0$, that implies $o_e\ell_e(o_e) > 0$).

By upper bounding all the weights w_is with o_e (we are allowed to do this since $w_i \leq o_e$ for any $i \in N$ and $e \in E$ such that $e \in \sigma_i^*$), by multiplying each constraint in (4.4) by w_i, and by summing all the resulting constraints in (4.4), we obtain the following relaxation of $\mathsf{LP1}$:

$$\begin{aligned} \mathsf{LP2}: \quad \max \quad & \sum_{e\in E} \alpha_e k_e \ell_e(k_e) \\ s.t. \quad & \sum_{e\in E} \alpha_e \beta_{\mathsf{W}}(\mathsf{PoA}_\varepsilon, k_e, o_e, \ell_e) \geq 0 && (4.6) \\ & \sum_{e\in E} \alpha_e o_e \ell_e(o_e) = 1 && (4.7) \end{aligned}$$

$$\alpha_e \geq 0, \quad \forall e \in E,$$

where $\beta_{\mathsf{W}}(\mathsf{PoA}_\varepsilon, k_e, o_e, \ell_e)$ is defined as in (4.1). Indeed, we have that

$$\sum_{i\in N} w_i \sum_{e\in\sigma_i} \alpha_e \ell_e(k_e) = \sum_{e\in E} \sum_{i\in N: e\in\sigma_i} \alpha_e w_i \ell_e(k_e) = \sum_{e\in E} \alpha_e k_e \ell_e(k_e)$$

and

$$\begin{aligned}\sum_{i\in N} w_i \sum_{e\in\sigma_i} \alpha_e \ell_e(k_e + w_i) &\leq \sum_{i\in N} w_i \sum_{e\in\sigma_i^*} \alpha_e \ell_e(k_e + o_e) \\ &= \sum_{e\in E} \sum_{i\in N: e\in\sigma_i^*} \alpha_e w_i \ell_e(k_e + o_e) \\ &= \sum_{e\in E} \alpha_e o_e \ell_e(k_e + o_e).\end{aligned}$$

By putting the above inequalities together, we get

$$\sum_{e\in E} \alpha_e k_e \ell_e(k_e) \leq (1+\varepsilon) \sum_{e\in E} \alpha_e o_e \ell_e(k_e + o_e),$$

that is equivalent to constraint (4.6). Consider the dual program of $\mathsf{LP2}$ in the dual variables x and γ (respectively associated to the dual constraints (4.6) and (4.7)):

$$\begin{aligned}\mathsf{DLP}: \quad & \min \quad \gamma \\ & s.t. \quad \gamma \cdot o_e \ell_e(o_e) \geq k_e \ell_e(k_e) + x \cdot \beta_{\mathsf{W}}(\mathsf{PoA}_\varepsilon, k_e, o_e, \ell_e), \quad \forall e \in E \\ & \qquad\; x \geq 0, \gamma \in \mathbb{R}.\end{aligned} \tag{4.8}$$

In DLP, one can assume that $x \geq 1$, otherwise, if $x < 1$ and $o_e = 0$ for some $e \in E$, the dual constraint in (4.8) related to e is not feasible. Now, one can assume that $o_e > 0$, since the constraint (4.8) corresponding to resource e is always verified if $o_e = 0$, so we can ignore this constraint. Let γ^* be the optimal value of DLP. By the Weak Duality Theorem, we get $\gamma^* \geq \mathsf{PoA}_\varepsilon(\mathcal{G})$. By (4.8) and the optimality of γ^*, we get[2]

$$\gamma^* = \min_{x\geq 1} \max_{e\in E} \gamma_{\mathsf{W}}(\mathsf{PoA}_\varepsilon, x, k_e, o_e, \ell_e) \leq \inf_{x\geq 1} \sup_{\substack{f\in\mathcal{C}(\mathcal{G}),\\ k>0, o>0}} \gamma_{\mathsf{W}}(\mathsf{EM}, x, k, o, f) = \gamma_{\mathsf{W}}(\mathsf{PoA}_\varepsilon, \mathcal{G}), \tag{4.9}$$

thus showing the claim for $\mathsf{EM} = \mathsf{PoA}_\varepsilon$.

For $\mathsf{EM} = \mathsf{CR}^s_\varepsilon$, one can use a similar proof as in the previous case. Let $\mathsf{CG} \in \mathcal{G}$ be an arbitrary congestion game, $\boldsymbol{\tau} = (\boldsymbol{\sigma}^0, \boldsymbol{\sigma}^1, \ldots, \boldsymbol{\sigma}^n)$ be an ε-approximate one-round walk involving selfish players, where $\boldsymbol{\sigma} := \boldsymbol{\sigma}^n$ is the strategy profile generated by $\boldsymbol{\tau}$, and let $\boldsymbol{\sigma}^*$ be an optimal strategy profile of CG. We can consider the linear program

[2] By exploiting the continuity of function $-kf(k)$ with respect to $k \geq 0$, we have that the supremum (appearing in the inequalities) can be equivalently taken over $k > 0$ instead of $k \geq 0$, without reducing its value.

LP1, but with constraint

$$\sum_{e\in\sigma_i}\alpha_e\ell_e(k_e(\boldsymbol{\sigma}^i))\leq(1+\varepsilon)\sum_{e\in\sigma_i^*}\alpha_e\ell_e(k_e(\boldsymbol{\sigma}^{i-1})+w_i)\tag{4.10}$$

in place of (4.4), since constraint (4.10) is necessarily satisfied by the greedy choice of each selfish player $i\in N$ (when $\alpha_e=1$ for any $e\in E$); thus, by exploiting the same arguments as in the previous case, the maximum value of LP1 is an upper bound on $\mathsf{CR}^s_\varepsilon(\mathsf{CG})$.

We have that

$$\begin{aligned}
&\sum_{i\in N}w_i\sum_{e\in\sigma_i}\alpha_e\ell_e(k_e(\boldsymbol{\sigma}^i))\\
&=\sum_{e\in E}\sum_{i:e\in\sigma_i}w_i\alpha_e\ell_e(k_e(\boldsymbol{\sigma}^i))\\
&=\sum_{e\in E}\sum_{i:e\in\sigma_i}\overbrace{\left(\int_0^{k_e}\chi_{[k_e(\boldsymbol{\sigma}^{i-1}),k_e(\boldsymbol{\sigma}^i)]}(t)dt\right)}^{w_i}\alpha_e\ell_e(k_e(\boldsymbol{\sigma}^i))\\
&=\sum_{e\in E}\alpha_e\int_0^{k_e}\left(\sum_{i:e\in\sigma_i}\ell_e(k_e(\boldsymbol{\sigma}^i))\chi_{[k_e(\boldsymbol{\sigma}^{i-1}),k_e(\boldsymbol{\sigma}^i)]}(t)\right)dt\\
&=\sum_{e\in E}\alpha_e\int_0^{k_e}\left(\sum_{i\in N}\ell_e(k_e(\boldsymbol{\sigma}^i))\chi_{[k_e(\boldsymbol{\sigma}^{i-1}),k_e(\boldsymbol{\sigma}^i)]}(t)\right)dt &(4.11)\\
&\geq\sum_{e\in E}\alpha_e\int_0^{k_e}\left(\sum_{i\in N}\ell_e(t)\chi_{[k_e(\boldsymbol{\sigma}^{i-1}),k_e(\boldsymbol{\sigma}^i)]}(t)\right)dt\\
&=\sum_{e\in E}\alpha_e\int_0^{k_e}\ell_e(t)dt, &(4.12)
\end{aligned}$$

where χ denotes the indicator function, and (4.11) holds since $\chi_{[k_e(\boldsymbol{\sigma}^{i-1}),k_e(\boldsymbol{\sigma}^i)]}(t)$ is null if $e\notin\sigma_i$ (as $k_e(\boldsymbol{\sigma}^{i-1})=k_e(\boldsymbol{\sigma}^i)$ in such case); furthermore, we have that

$$\sum_{i\in N}w_i\sum_{e\in\sigma_i^*}\alpha_e\ell_e(k_e(\boldsymbol{\sigma}^{i-1})+w_i)\leq\sum_{i\in N}w_i\sum_{e\in\sigma_i^*}\alpha_e\ell_e(k_e+o_e)=\sum_{e\in E}\alpha_e o_e\ell_e(k_e+o_e).\tag{4.13}$$

Thus, by (4.10), (4.12) and (4.13), we get

$$\begin{aligned}
\sum_{e\in E}\alpha_e\int_{t=0}^{k_e}\ell_e(t)dt&\leq\sum_{i\in N}w_i\sum_{e\in\sigma_i}\alpha_e\ell_e(k_e(\boldsymbol{\sigma}^i))\\
&\leq\sum_{i\in N}w_i(1+\varepsilon)\sum_{e\in\sigma_i^*}\alpha_e\ell_e(k_e(\boldsymbol{\sigma}^{i-1})+w_i)\\
&\leq\sum_{e\in E}\alpha_e(1+\varepsilon)o_e\ell_e(k_e+o_e).
\end{aligned}$$

Such inequality is equivalent to constraint (4.6), but with $\mathsf{CR}^s_\varepsilon$ in place of PoA_ε; thus, we can continue the analysis from a linear program analogue to $\mathsf{LP2}$, and by proceeding as in the case of the approximate price of anarchy, we can show the claim for $\mathsf{EM} = \mathsf{CR}^s_\varepsilon$.

Finally, we consider the case $\mathsf{EM} = \mathsf{CR}^c_\varepsilon$. Again, let $\mathsf{CG} \in \mathcal{G}$ be an arbitrary congestion game, $\boldsymbol{\tau} = (\boldsymbol{\sigma}^0, \boldsymbol{\sigma}^1, \ldots, \boldsymbol{\sigma}^n)$ be an ε-approximate one-round walk involving cooperative players generating strategy profile $\boldsymbol{\sigma} = \boldsymbol{\sigma}^n$, and $\boldsymbol{\sigma}^*$ be an optimal strategy profile of CG. Let $\hat{\ell}_e$ denote the function defined as $\hat{\ell}_e(x) := x\ell_e(x)$ for any $x \geq 0$. By the hypothesis, $\hat{\ell}_e$ is a convex function for any $e \in E$. Again, we can consider the linear program $\mathsf{LP1}$, but with constraint

$$\sum_{e\in E} \alpha_e(\hat{\ell}_e(k_e(\boldsymbol{\sigma}^i)) - \hat{\ell}_e(k_e(\boldsymbol{\sigma}^{i-1}))) \leq (1+\varepsilon)\sum_{e\in E} \alpha_e(\hat{\ell}_e(k_e(\boldsymbol{\sigma}^{i-1}_{-i}, \sigma_i^*)) - \hat{\ell}_e(k_e(\boldsymbol{\sigma}^{i-1}))) \tag{4.14}$$

in place of (4.4), as (4.14) is necessarily satisfied by the greedy choice of each cooperative player $i \in N$ (when $\alpha_e = 1$ for any $e \in E$); thus, the maximum value of $\mathsf{LP1}$ is an upper bound on $\mathsf{CR}^c_\varepsilon(\mathsf{CG})$.

We have that

$$\begin{aligned}
&\sum_{e\in E} \alpha_e k_e \ell_e(k_e) \\
&= \sum_{e\in E} \alpha_e \hat{\ell}_e(k_e) \\
&= \sum_{e\in E} \alpha_e \sum_{i\in N} (\hat{\ell}_e(k_e(\boldsymbol{\sigma}^i)) - \hat{\ell}_e(k_e(\boldsymbol{\sigma}^{i-1}))) \\
&= \sum_{i\in N}\sum_{e\in E} \alpha_e (\hat{\ell}_e(k_e(\boldsymbol{\sigma}^i)) - \hat{\ell}_e(k_e(\boldsymbol{\sigma}^{i-1}))) \\
&\leq (1+\varepsilon)\sum_{e\in E} \alpha_e(\hat{\ell}_e(k_e(\boldsymbol{\sigma}^{i-1}_{-i}, \sigma_i^*)) - \hat{\ell}_e(k_e(\boldsymbol{\sigma}^{i-1}))) && (4.15) \\
&= \sum_{i\in N}(1+\varepsilon)\sum_{e\in E} \alpha_e(\hat{\ell}_e(k_e(\boldsymbol{\sigma}^{i-1}) + w_i \cdot \chi_{\sigma_i^*}(e)) - \hat{\ell}_e(k_e(\boldsymbol{\sigma}^{i-1}))) \\
&\leq \sum_{i\in N}(1+\varepsilon)\sum_{e\in \sigma_i^*} \alpha_e(\hat{\ell}_e(k_e(\boldsymbol{\sigma}^{i-1}) + w_i) - \hat{\ell}_e(k_e(\boldsymbol{\sigma}^{i-1}))) \\
&= \sum_{e\in E} \alpha_e \sum_{i\in N: e\in\sigma_i^*} (1+\varepsilon)\left(\hat{\ell}_e(k_e(\boldsymbol{\sigma}^{i-1}) + w_i) - \hat{\ell}_e(k_e(\boldsymbol{\sigma}^{i-1}))\right) \\
&\leq \sum_{e\in E} \alpha_e \sum_{i\in N: e\in\sigma_i^*} (1+\varepsilon)\left(\hat{\ell}_e(k_e(\boldsymbol{\sigma}) + w_i) - \hat{\ell}_e(k_e(\boldsymbol{\sigma}))\right) && (4.16) \\
&\leq \sum_{e\in E} \alpha_e \sum_{\substack{i\in N:\\ e\in\sigma_i^*}} (1+\varepsilon)\left(\hat{\ell}_e\left(k_e(\boldsymbol{\sigma}) + \sum_{\substack{h\in[i]:\\ e\in\sigma_h^*}} w_h\right) - \hat{\ell}_e\left(k_e(\boldsymbol{\sigma}) + \sum_{\substack{h\in[i-1]:\\ e\in\sigma_h^*}} w_h\right)\right) && (4.17)
\end{aligned}$$

$$= \sum_{e\in E} \alpha_e(1+\varepsilon)\left(\hat{\ell}_e\left(k_e(\boldsymbol{\sigma}) + \sum_{h\in N: e\in\sigma_h^*} w_h\right) - \hat{\ell}_e(k_e(\boldsymbol{\sigma}))\right) \tag{4.18}$$

$$= \sum_{e\in E} \alpha_e(1+\varepsilon)\left((k_e+o_e)\ell_e(k_e+o_e) - k_e\ell_e(k_e)\right),$$

where (4.15) comes from (4.14), (4.16) and (4.17) hold because of the convexity of the functions $\hat{\ell}_e$s (as $F(x+y+z) - F(x+y) \geq F(y+z) - F(y)$ holds for any non-negative and non-decreasing convex function F, and for any $x,y,z \geq 0$), and (4.18) holds since the inner sum of (4.17) is telescoping. Thus, inequality

$$\sum_{e\in E} \alpha_e k_e \ell_e(k_e) \leq \sum_{e\in E} \alpha_e(1+\varepsilon)((k_e+o_e)\ell_e(k_e+o_e) - k_e\ell_e(k_e))$$

holds. Again, such inequality is equivalent to constraint (4.6), but with $\mathsf{CR}^c_\varepsilon$ in place of PoA_ε; thus, we can continue the analysis from a linear program analogue to LP2, and by proceeding as in the case of the approximate price of anarchy, we can show the claim for $\mathsf{EM} = \mathsf{CR}^c_\varepsilon$. □

Remark 4.1 (Robustness of the Approximate Price of Anarchy). The efficiency metric of the price of anarchy can be equivalently redefined to measure the performance with respect to the probabilistic generalizations of pure Nash equilibria defined in Section 2.3 (i.e., mixed Nash equilibria, correlated and coarse-correlated equilibria). By exploiting the results of Roughgarden [158] and Bilò [42], one can show that the upper bounds on the approximate price of anarchy provided in Theorem 4.1 continue to hold even with respect to the generalized notions of equilibrium.

Lemma 4.1. *Let $\mathcal{G}$ be a class of weighted congestion games. For each $M < \gamma_W(\mathsf{EM}, \mathcal{G})$, with $\mathsf{EM} \in \{\mathsf{PoA}_\varepsilon, \mathsf{CR}^s_\varepsilon\}$, one of the following cases holds:*

- **Case 1:** *there exists a non-constant latency function $f \in \mathcal{C}(\mathcal{G})$ and two real numbers $k, o > 0$ such that*

$$M < \frac{kf(k)}{of(o)} \tag{4.19}$$

and $\beta_W(\mathsf{EM}, k, o, f) \geq 0$.

- **Case 2:** *there exist two latency functions $f_1, f_2 \in \mathcal{C}(\mathcal{G})$ and four real numbers $k_1, k_2, o_1, o_2 > 0$ such that*

$$M < \frac{\alpha_1 k_1 f_1(k_1) + \alpha_2 k_2 f_2(k_2)}{\alpha_1 o_1 f_1(o_1) + \alpha_2 o_2 f_2(o_2)}, \tag{4.20}$$

where $\alpha_1 := \beta_W(\mathsf{EM}, k_2, o_2, f_2) > 0$ and $\alpha_2 := -\beta_W(\mathsf{EM}, k_1, o_1, f_1) > 0$; furthermore, if $\mathsf{EM} = \mathsf{PoA}_\varepsilon$, f_1 and f_2 can be chosen as non-constant latency functions.

This fact holds for $\mathsf{EM} = \mathsf{CR}^c_\varepsilon$ if the latency functions of $\mathcal{C}(\mathcal{G})$ are semi-convex.

Proof (Sketch). The claim of Lemma 4.1 is obtained by inverting the proof arguments of Theorem 4.1 as follows:

- We start from the upper bound $\gamma_{\mathsf{W}}(\mathsf{EM},\mathcal{G})$ obtained in Theorem 4.1.
- We derive a linear program $\overline{\mathsf{DLP}}$ similar to the dual program used in Theorem 4.1, but with two constraints only (except to those imposing the non-negativity of the variables), and whose optimal value is higher than M. In particular, we show that there exist two triples of parameters $T_1=(k_1,o_1,f_1)$ and $T_2=(k_2,o_2,f_2)$, with $k_1,k_2,o_1,o_2>0$ and $f_1,f_2\in\mathcal{C}$, such that $\beta_{\mathsf{W}}(\mathsf{EM},k_1,o_1,f_1)<0$, $\beta_{\mathsf{W}}(\mathsf{EM},k_2,o_2,f_2)\geq 0$, and such that the two dual constraints of $\overline{\mathsf{DLP}}$ (which are defined as in (4.8)) are parametrized by T_1 and T_2, respectively.
- The dual of $\overline{\mathsf{DLP}}$, denoted as $\overline{\mathsf{LP}}$, is similar to the linear program $\mathsf{LP2}$ used in Theorem 4.1, but with two constraints and two variables, and has the same optimal value (higher than M) as in $\overline{\mathsf{DLP}}$ (by the Strong Duality Theorem); in particular, the new linear program is explicitly defined as follows:

$$\begin{aligned}\overline{\mathsf{LP}}:\quad \max\quad & \alpha_1 k_1 f_1(k_1)+\alpha_2 k_2 f_2(k_2)\\ s.t.\quad & \alpha_1\beta_{\mathsf{W}}(\mathsf{EM},k_1,k_1,f_1)+\alpha_2\beta_{\mathsf{W}}(\mathsf{EM},k_2,o_2,f_2)\geq 0\\ & \alpha_1 o_1 f_1(o_1)+\alpha_2 o_2 f_2(o_2)=1\\ & \alpha_1,\alpha_2\geq 0.\end{aligned}$$

- Finally, the claim of Lemma 4.1 is obtained by characterizing the optimal solution of $\overline{\mathsf{LP}}$. □

Remark 4.2. By exploiting Lemma 4.1, we have that, for any choice of the input parameters k,o,f (resp. k_1,k_2,o_1,o_2,f_1,f_2) such that $\beta_{\mathsf{W}}(\mathsf{EM},k,o,f)\geq 0$ (resp. $\beta_{\mathsf{W}}(\mathsf{EM},k_2,o_2,f_2)>0$ and $-\beta_{\mathsf{W}}(\mathsf{EM},k_1,o_1,f_1)>0$), the right hand-part of (4.19) (resp. (4.20)) is a lower bound for $\gamma_{\mathsf{W}}(\mathsf{EM},\mathcal{G})$; we also observe that, by definition of $\gamma_{\mathsf{W}}(\mathsf{EM},\mathcal{G})$, $\sup_{f\in\mathcal{C}(\mathcal{G}),k,o>0}\gamma_{\mathsf{W}}(\mathsf{EM},x,k,o,f)$ is an upper bound on $\gamma_{\mathsf{W}}(\mathsf{EM},\mathcal{G})$ for any $x\geq 1$. Thus, if such upper bound (for a suitable choice of x) is equal to the former lower bound (for a suitable choice of the input parameters), we necessarily have that they are both equal to $\gamma_{\mathsf{W}}(\mathsf{EM},\mathcal{G})$.

Remark 4.3. In Lemma 4.1, if $\mathcal{C}(\mathcal{G})$ is closed under abscissa scaling, we can assume without loss of generality that $o_j=1$ for each $j\in[2]$ (remove index j if the first case of Lemma 4.1 is verified). Indeed, if it is not the case, let $\hat{f}_j\in\mathcal{C}(\mathcal{G})$ be the latency function such that $\hat{f}_j(k):=f_j(o_jk)$ for any $k\geq 0$. If we consider tuple $(k_j/o_j,1,\hat{f}_j)$ in place of tuple (k_j,o_j,f_j) for each $j\in[2]$, the claim of Lemma 4.1 follows as well. Analogously, if $\mathcal{C}(\mathcal{G})$ is closed under abscissa scaling, the supremum appearing in the definition of $\gamma_{\mathsf{W}}(\mathsf{EM},\mathcal{C})$ does not decrease when assuming $o=1$; thus, also Theorem 4.1 holds under this assumption.

Example 4.1. To illustrate the claims of Theorem 4.1 and Lemma 4.1 with a more concrete example, we consider the problem of evaluating the competitive ratio of exact one-round walks involving selfish players for affine weighted congestion games. Because of Theorem 4.1 and Remark 4.3, we get that:

$$\gamma_{\mathsf{W}}(\mathsf{CR}_0^s,\mathcal{P}(1))=\inf_{x\geq 1}\sup_{f\in\mathcal{P}(1),k\geq 0}\left(kf(k)+x\left(f(k+1)-\int_0^k f(t)dt\right)\right)$$

$$
\begin{aligned}
&= \inf_{x\geq 1} \sup_{\substack{(\alpha_0,\alpha_1)>0\\ k>0}} \left(\sum_{d\in\{0,1\}} \alpha_d k^{d+1} + x\left(\sum_{d\in\{0,1\}} \alpha_d (k+1)^d - \sum_{d\in\{0,1\}} \alpha_d \frac{k^{d+1}}{d+1} \right)\right) \\
&= \inf_{x\geq 1} \sup_{d\in\{0,1\},k>0} \left(k^{d+1} + x\left((k+1)^d - \frac{k^{d+1}}{d+1} \right)\right) \\
&= \min_{x>2} \sup_{k>0} \left(k^2 + x\left(k+1 - \frac{k^2}{2} \right)\right) \\
&\leq \gamma\left(\mathsf{CR}_0^s, \overbrace{\frac{2\sqrt{3}+6}{3}}^{x}, \overbrace{\frac{\sqrt{3}+3}{\sqrt{3}}}^{k}, \overbrace{1}^{o}, f: f(t)=t \right) \qquad (4.21) \\
&= 2\sqrt{3}+4,
\end{aligned}
$$

thus obtaining the same upper bound of Christodoulou et al. [63]. Relatively to Lemma 4.1, we have that the first case is verified. In particular, by setting $k = \frac{\sqrt{3}+3}{\sqrt{3}}$, $o = 1$ and f in such a way that $f(t) = t$ for any $t \geq 0$ (i.e. the parameters k, o, f used in (4.21)), we get that the second inequality of the first case of Lemma 4.1 is tight (i.e., $\beta_{\mathsf{W}}(\mathsf{CR}_\varepsilon^s, k, o, f) = 0$), and $\frac{kf(k)}{of(o)}$ is equal to $2\sqrt{3}+4$. Thus, by Remark 4.2, we necessarily have that $\gamma_{\mathsf{W}}(\mathsf{CR}_0^s, \mathcal{P}(1)) = 2\sqrt{3}+4$.

4.2 Load Balancing Instances as Tight Lower Bounds

In this section, by using the characterization provided in Lemma 4.1, we show that the upper bounds provided in Theorem 4.1 are tight under mild assumptions on the considered latency functions, even for load balancing games.

Theorem 4.2. *Let $\mathcal{C}$ be a class of latency functions that is closed under ordinate and abscissa scaling. Then,* $\gamma_{\mathsf{W}}(\mathsf{PoA}_\varepsilon, \mathsf{W}(\mathcal{C})) = \mathsf{PoA}_\varepsilon(\mathsf{W}(\mathcal{C})) = \mathsf{PoA}_\varepsilon(\mathsf{WLB}(\mathcal{C}))$. *If all latency functions of $\mathcal{C}$ (except for the constant ones) are unbounded, we get* $\gamma_{\mathsf{W}}(\mathsf{PoA}_\varepsilon, \mathsf{W}(\mathcal{C})) = \mathsf{PoA}_\varepsilon(\mathsf{W}(\mathcal{C})) = \mathsf{PoA}_\varepsilon(\mathsf{WSLB}(\mathcal{C}))$.

Proof (Sketch). First of all, we consider the case in which all latency functions, except for the constant ones, are unbounded. Fix $M < \gamma_{\mathsf{W}}(\mathsf{PoA}_\varepsilon, \mathsf{W}(\mathcal{C}))$. In the proof, we will show that there exists a load balancing instance $\mathsf{LB} \in \mathsf{WSLB}(\mathcal{C})$ such that $\mathsf{PoA}_\varepsilon(\mathsf{LB}) > M$. By Lemma 4.1, this will imply that

$$M < \mathsf{PoA}_\varepsilon(\mathsf{LB}) \leq \mathsf{PoA}_\varepsilon(\mathsf{W}(\mathcal{C})) \leq \gamma_{\mathsf{W}}(\mathsf{PoA}_\varepsilon, \mathsf{W}(\mathcal{C})),$$

and by the arbitrariness of $M < \gamma_{\mathsf{W}}(\mathsf{PoA}_\varepsilon, \mathsf{W}(\mathcal{C}))$, this fact will show the claim. We make use of a multi-graph to represent a load balancing game together with two special strategy profiles. This multi-graph is denoted as *load balancing graph*, and is defined as follows: the nodes are all the resources in E and each player is associated to a weighted edge (e_1, e_2, w), where $\{e_1\}$ is denoted as her first strategy, $\{e_2\}$ is her second strategy, and w is her weight. With a little abuse of notation, any load balancing graph will be also used to denote the corresponding load balancing game.

If Case 2 of Lemma 4.1 holds (with respect to $\mathcal{C}$), let $k_1, k_2, o_1, o_2, f_1, f_2, \alpha_1 = \beta_W(\mathsf{PoA}_\varepsilon, k_2, o_2, f_2), \alpha_2 = -\beta_W(\mathsf{PoA}_\varepsilon, k_1, o_1, f_1)$ be the parameters considered in the claim of Lemma 4.1. By Lemma 4.1, f_1 and f_2 can be chosen among the non-constant latencies of $\mathcal{C}$, thus, by hypothesis, they are unbounded. Furthermore, by Remark 4.3, we can assume without loss of generality that $o_1 = o_2 = 1$.

Let $s, n \in \mathbb{N}$ be two arbitrary positive integers. Consider a load balancing graph $\mathsf{LB}_{s,n}(k_1, k_2, f_1, f_2)$ yielded by a directed n-ary tree[3], organized in $2s$ levels, numbered from 1 to $2s$, and whose edges are oriented from the root to the leaves, with the addition of n self-loops on the nodes of level $2s$. The weight w_i of a player associated to an edge outgoing from a node at level $i \in [s]$ (resp. $i \in [2s]_{s+1}$) is equal to $(k_1/n)^i$ (resp. $(k_1/n)^s(k_2/n)^{i-s}$). For $i, j \in [2]$, define

$$\theta_{i,j} := \frac{f_i(k_i)}{(1+\varepsilon)f_j(k_j+1)}$$

and $\theta_i := \theta_{i,i}$. Each resource at level i has latency

$$g_i(x) := \begin{cases} \theta_1^{i-1} f_1\left(\left(\frac{n}{k_1}\right)^{i-1} x\right) & \text{if } i \in [s], \\ \theta_1^{s-1}\theta_{1,2}\theta_2^{i-s-1} f_2\left(\left(\frac{n}{k_1}\right)^s \left(\frac{n}{k_2}\right)^{i-s-1} x\right) & \text{otherwise.} \end{cases}$$

See Figure 4.1 for an example. Let $\boldsymbol{\sigma}_{s,n}$ be the strategy profile in which all players select their first strategy. By standard calculations, one can show the following lemma:

Lemma 4.2. *For any integer $s \geq 2$, there exists $n(s) \in \mathbb{N}$, such that $\boldsymbol{\sigma}_{s,n}$ is an ε-approximate pure Nash equilibrium of $\mathsf{LB}_{s,n}(k_1, k_2, f_1, f_2)$ for any $n \geq n(s)$.*

The proof of the above lemma is omitted, and we only give a sketch of the proof. We consider an arbitrary player whose first strategy is a resource from a level i. Since the game is symmetric, we have that: (i) if $i \in [2s-1]$ and the player deviates in favor of a resource from level $j = i+1$, her cost decreases exactly of a factor $1+\varepsilon$; (ii) if $i \in [2s]$ and the player deviates in favor of a resource from level $j \leq i$, her cost does not decrease; (iii) if $i \in [2s-2]$ and the player deviates in favor of a resource from level $j > i+1$, if n is sufficiently large, her cost does not decrease. In any case, the cost of the considered player, when she deviates, cannot decrease of a factor higher than $1+\varepsilon$, thus $\boldsymbol{\sigma}_{s,n}$ is a pure Nash equilibrium.

In the remainder of the proof, with the aim of estimating the approximate price of anarchy of $\mathsf{LB}_{s,n}(k_1, k_2, f_1, f_2)$, we compare the social value of the ε-pure Nash

[3] With a little abuse of notation, the value of n considered in this load-balancing instance does not represent the total number of players, but the number of players selecting each resource when playing their first strategy.

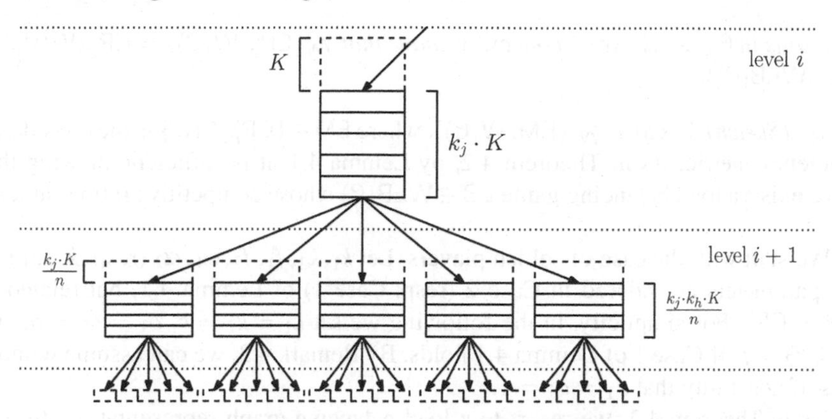

Fig. 4.1 A node/resource at some level $i \in [2s-2]_2$ of the load balancing graph $\mathsf{LB}_{s,n}(k_1,k_2,f_1,f_2)$ with $n=5$, and all the nodes/resources at level $i+1$ connected to it. Each non-dashed (resp. dashed) rectangle on some node/resource represents a player selecting that resource in $\boldsymbol{\sigma}_{s,n}$ (resp. $\boldsymbol{\sigma}^*_{s,n}$). The congestion of each resource in $\boldsymbol{\sigma}_{s,n}$ or $\boldsymbol{\sigma}^*_{s,n}$ is represented accordingly. Let K be the congestion of the resources at level i in $\boldsymbol{\sigma}^*_{s,n}$. Then $K \cdot k_j$ is the congestion of the resources at level i in $\boldsymbol{\sigma}_{s,n}$, $K \cdot k_j/n$ is the congestion of the resources at level $i+1$ in $\boldsymbol{\sigma}^*_{s,n}$, and $K \cdot k_j \cdot k_h/n$ is the congestion of the resources at level $i+1$ in $\boldsymbol{\sigma}_{s,n}$, where $j=h=1$ if $i \in [s-1]$, $j=1$ and $h=2$ if $i=s$, $j=h=2$ otherwise.

equilibrium $\boldsymbol{\sigma}_{s,n}$, with that of the strategy profile $\boldsymbol{\sigma}^*_{s,n}$ in which all players select their second strategy. By exploiting the definition of α_1 and α_2 given in Lemma 4.1, by standard (but long) calculations one can show that the quantity $\frac{\mathsf{SUM}(\boldsymbol{\sigma}_{s,n})}{\mathsf{SUM}(\boldsymbol{\sigma}^*_{s,n})}$ approaches the value $\frac{\alpha_1 k_1 f_1(k_1)+\alpha_2 k_2 f_2(k_2)}{\alpha_1 o_1 f_1(o_1)+\alpha_2 o_2 f_2(o_2)}$, as n and s tend to ∞. Thus, by Lemma 4.1, we get $\mathsf{PoA}_\varepsilon(\mathsf{LB}_{s,n}(k_1,k_2,f_1,f_2)) > M$ for sufficiently large n and s.

Now, suppose that Case 1 of Lemma 4.1 holds, and let k,o,f be the parameters considered in this case. As done previously, by Remark 4.3, we can assume without loss of generality that $o=1$. The load balancing instance we consider here is $\mathsf{LB}_{s,n}(k,k,f,f)$, and the strategy profiles $\boldsymbol{\sigma}_{s,n}$ and $\boldsymbol{\sigma}^*_{s,n}$ are defined as in the previous part of the proof. By using the same proof arguments of the above case, we get $\mathsf{PoA}_\varepsilon(\mathsf{LB}_{s,n}(k,k,f,f)) > M$ for sufficiently large n and s.

Finally, if there is some non-constant latency function in $\mathcal{C}(\mathcal{G})$ that is not unbounded, we can consider a non-symmetric load balancing game based on the load balancing graph $\mathsf{LB}_{s,n}(k_1,k_2,f_1,f_2)$ defined above with $n=1$ (i.e., the load balancing graph is a directed path), but where each player can select among her first and second strategy only, and based on the parameters k_1,k_2,f_1,f_2 or k,f derived according to Lemma 4.1. Also in this case, by using the same proof arguments of the above cases, we get $\mathsf{PoA}_\varepsilon(\mathsf{LB}_{s,n}(k_1,k_2,f_1,f_2)) > M$ for a sufficiently large s. □

Now, we prove that no improvements are possible for approximate one-round walks when restricting to load balancing games.

Theorem 4.3. *Let $\mathcal{C}$ be a class of latency functions that is closed under ordinate and abscissa scaling. Then,* $\gamma_{\mathsf{W}}(\mathsf{CR}^s_\varepsilon,\mathsf{W}(\mathcal{C})) = \mathsf{CR}^s_\varepsilon(\mathsf{W}(\mathcal{C})) = \mathsf{CR}^s_\varepsilon(\mathsf{WLB}(\mathcal{C}))$. *If all* *functions in $\mathcal{C}$ are also semi-convex, we have that* $\gamma_{\mathsf{W}}(\mathsf{CR}^c_\varepsilon, \mathsf{W}(\mathcal{C})) = \mathsf{CR}^c_\varepsilon(\mathsf{W}(\mathcal{C})) = \mathsf{CR}^c_\varepsilon(\mathsf{WLB}(\mathcal{C}))$.

Proof (Sketch). Fix $M < \gamma_{\mathsf{W}}(\mathsf{EM}, \mathsf{W}(\mathcal{C}))$, where $\mathsf{EM} \in \{\mathsf{CR}^s_\varepsilon, \mathsf{CR}^c_\varepsilon\}$ is the considered efficiency metric. As in Theorem 4.2, by Lemma 4.1, it is sufficient showing that there exists a load balancing game $\mathsf{LB} \in \mathsf{WLB}(\mathcal{C})$ whose competitive ratio is at least M.

We start with the case of selfish players. Let $k_1, k_2, f_1, f_2, \alpha_1, \alpha_2$ (resp. k, o, f) be the parameters considered in Case 2 (resp. Case 1) of Lemma 4.1, but related to $\mathsf{EM} = \mathsf{CR}^s_\varepsilon$. For simplicity, in the following we set $k_1 = k_2 = k$, $o_1 = o_2 = o$, and $f_1 = f_2 = f$, if Case 1 of Lemma 4.1 holds. By Remark 4.3, we can assume without loss of generality that $o_1 = o_2 = 1$.

As in Theorem 4.2, we resort to a load balancing graph representation. In particular, we extend the load balancing graph $\mathsf{LB}_{s,n}(k_1, k_2, f_1, f_2)$ defined in the proof of Theorem 4.2 as follows. Denote as $i(v)$ the level of resource v. For each node u in the load balancing graph, consider an arbitrary enumeration of all the n outgoing edges of u. Since each node (but the root) has a unique incoming edge, we denote by $h(v) \in [n]$ the label/position associated to the unique edge entering v in the given ordering, so that, for any node u (but the leaves), all the n edges of type (u, v) are associated to a label $h(v) \in [n]$.

For $i, j \in [2]$ and $h \in [n]$, define

$$\theta_{i,j}(h) := \frac{f_i\left(\frac{hk_i}{n}\right)}{(1+\varepsilon) f_j(k_j + 1)}$$

and $\theta_i(h) := \theta_{i,i}(h)$. Resource v has latency function

$$g_v(x) := \begin{cases} f_1(x) & \text{if } i(v) = 1, \\ \underbrace{\theta_1(h(v)) A_u}_{A_v} f_1\left(\left(\frac{n}{k_1}\right)^{i(v)-1} x\right) & \text{if } i(v) \in [s]_2, \\ \underbrace{\theta_{1,2}(h(v)) A_u}_{A_v} f_2\left(\left(\frac{n}{k_1}\right)^{s} x\right) & \text{if } i(v) = s+1, \\ \underbrace{\theta_2(h(v)) A_u}_{A_v} f_2\left(\left(\frac{n}{k_1}\right)^{s} \left(\frac{n}{k_2}\right)^{i(v)-s-1} x\right) & \text{if } i(v) \in [2s]_{s+2}, \end{cases}$$

where (u, v) denotes the unique incoming edge of v and A_v is recursively defined on the basis of A_u as follows:

$$A_v := \begin{cases} 1 & \text{if } i(v) = 1, \\ \theta_1(h(v)) & \text{if } i(v) \in [s]_2, \\ \theta_{1,2}(h(v)) A_u & \text{if } i(v) = s+1, \\ \theta_2(h(v)) A_u & \text{if } i(v) \in [2s]_{s+2}. \end{cases}$$

The weights of all players are defined as in Theorem 4.2.

Consider the online process $\boldsymbol{\tau}_{s,n}$ in which players enter the game in non-increasing order of level (with respect to their first strategy) and, within the same level, players are processed in non-decreasing order of position; equivalently, if two players i_u and i_v have, as their first strategies, some resources u and v respectively, such that either $i(u) > i(v)$, or $i(u) = i(v) \wedge h(u) < h(v)$, then player i_u is processed before player i_v in the online process $\boldsymbol{\tau}_{s,n}$. Let $\boldsymbol{\sigma}_{s,n}$ be strategy profile obtained at the end of the process, i.e., the strategy profile in which each player selects her first strategy. We have the following lemma:

Lemma 4.3. *The online process $\boldsymbol{\tau}_{s,n}$ is an ε-approximate one-round walk.*

The above lemma can be shown as follows. Let z be an arbitrary player whose first strategy is at level $i \in [2s-1]$; let u, v be the first and second strategy of z, respectively, and let $h := h(v)$ be the position/label associated to resource v. By construction of the online process $\boldsymbol{\tau}_{s,n}$, we have that, when z enters the game, there are $h-1$ players of weight w_i already assigned to resource u, and n players of weight w_{i+1} assigned to resource v. Let $\boldsymbol{\sigma}^u_{s,n}$ be the partial strategy profile obtained when z is processed according to $\boldsymbol{\tau}_{s,n}$ (i.e., only players preceding z have been already assigned), and z is assigned to her first strategy u; let $\boldsymbol{\sigma}^v_{s,n}$ be the partial strategy profile in which z, instead, is assigned to her second strategy v. As our choice of player z has been arbitrary, it is sufficient proving that $cost_z(\boldsymbol{\sigma}^u_{s,n}) \leq (1+\varepsilon)cost_z(\boldsymbol{\sigma}^v_{s,n})$ to show that $\boldsymbol{\tau}_{s,n}$ is an ε-approximate one-round walk. By using the recursive definitions of the latency functions g_i and g_{i+1} (involving the quantities A_u and A_v, respectively), and by using a similar approach as in case (i) of Lemma 4.2, one can easily show that $cost_z(\boldsymbol{\sigma}^u_{s,n}) = g_u(h \cdot w_i) = (1+\varepsilon)g_v(n \cdot w_{i+1} + w_i) = (1+\varepsilon)cost_z(\boldsymbol{\sigma}^v_{s,n})$, thus showing the claim of Lemma 4.3.

Now, let $\boldsymbol{\sigma}^*_{s,n}$ be the strategy profile in which all players select their second strategy. With the aim of estimating the competitive ratio of $\mathsf{LB}_{s,n}(k_1,k_2,f_1,f_2)$, we compare the social value of the strategy profile $\boldsymbol{\sigma}_{s,n}$ resulting from the ε-approximate one-round walk $\boldsymbol{\tau}_{s,n}$ with that of strategy profile $\boldsymbol{\sigma}^*_{s,n}$. As in Theorem 4.2, one can show that for sufficiently large n and s, $\mathsf{CR}^s_\varepsilon(\mathsf{LB}_{s,n}(k_1,k_2,f_1,f_2)) > M$.

For cooperative players, let $k_1,k_2,o_1,o_2,f_1,f_2,\alpha_1,\alpha_2$ be as in Lemma 4.1 (remove index $j \in [2]$ and set $\alpha = 1$ if Case 1 of Lemma 4.1 is verified). By Remark 4.3, we assume without loss of generality that $o_1 = o_2 = 1$. Consider a load balancing graph $\mathsf{LB}_{s,n}(k_1,k_2,f_1,f_2)$ as the one defined above, but with $n = 1$ and $\theta_{i,j}(1) := \frac{f_i(k_i)}{(1+\varepsilon)((k_j+1)f_j(k_j+1)-k_jf_j(k_j))}$. Let $\boldsymbol{\sigma}_{s,1}$, $\boldsymbol{\tau}_{s,1}$, and $\boldsymbol{\sigma}^*_{s,1}$ be defined as in the case of selfish players (with $n = 1$). Similarly as in the case of selfish players, one can show that $\boldsymbol{\tau}_{s,1}$ is an ε-approximate one-round walk (involving cooperative players) that generates strategy profile $\boldsymbol{\sigma}_{s,1}$, and we can show again that $\mathsf{CR}^c_\varepsilon(\mathsf{LB}_{s,n}(k_1,k_2,f_1,f_2)) > M$. □

Remark 4.4. Fix a metric $\mathsf{EM} \in \{\mathsf{PoA}_\varepsilon, \mathsf{CR}^s_\varepsilon, \mathsf{CR}^c_\varepsilon\}$, a class of latency functions $\mathcal{C}$, and let $(k_1,k_2,o_1,o_2,f_1,f_2)$ (resp. (k,o,f)) be a tuple such that the values α_1,α_2 considered in Case 2 of Lemma 4.1 are positive (resp. such that the inequality of Case 1 corresponding to the considered metric EM is satisfied, i.e.,

$\beta_{\mathsf{W}}(\mathsf{EM},k,o,f) \geq 0$). By Remark 4.3, we assume without loss of generality that $o_1 = o_2 = 1$ (resp. $o = 1$). By inspecting the proofs of Theorems 4.2 and 4.3, we observe that the set of load balancing instances of type $\mathsf{LB}_{s,n}(k_1,k_2,f_1,f_2)$ (resp. $\mathsf{LB}_{s,n}(k,k,f,f)$) guarantees a performance of at least $\frac{\alpha_1 k_1 f_1(k_1)+\alpha_2 k_2 f_2(k_2)}{\alpha_1 f(1)+\alpha_2 f(2)}$ (resp. $\frac{kf(k)}{f(1)}$) under the considered metric EM. Thus, in the case we are not able to quantify the exact value of $\gamma_{\mathsf{W}}(\mathsf{EM},\mathsf{W}(\mathcal{C}))$, we can still find a tuple (k_1,k_2,f_1,f_2) (resp. (k,f)) that leads to good (and possibly tight) lower bounds; in particular, the tightness can be shown by using the characterization provided in Remark 4.2.

4.3 Application to Polynomial Latency Functions

Consider the class $\mathcal{P}(d)$ of polynomials with non-negative coefficients and maximum degree d. By Lemma 4.1, $\gamma_{\mathsf{W}}(\mathsf{EM},\mathsf{W}(\mathcal{P}(d)))$ is an upper bound on the efficiency metrics $\mathsf{EM} \in \{\mathsf{PoA}_\varepsilon, \mathsf{CR}^s_\varepsilon, \mathsf{CR}^c_\varepsilon\}$ for weighted congestion games and load balancing games with polynomial latency functions of maximum degree d.

By exploiting the definition of $\gamma_{\mathsf{W}}(\mathsf{EM},\mathsf{W}(\mathcal{P}(d)))$ given at the beginning of this section, and by applying to these definitions similar arguments as those used in Aland et al. [7] and Christodoulou et al. [63], one can compute the exact value of $\gamma_{\mathsf{W}}(\mathsf{EM},\mathsf{W}(\mathcal{P}(d)))$ for any $\mathsf{EM} \in \{\mathsf{PoA}_\varepsilon, \mathsf{CR}^s_\varepsilon, \mathsf{CR}^c_\varepsilon\}$. In particular, we have that $\gamma_{\mathsf{W}}(\mathsf{EM},\mathsf{W}(\mathcal{P}(d)))$ is equal to $(\varphi_{\varepsilon,d})^d$, where $\varphi_{\varepsilon,d}$ is defined as the unique solution $k \geq 0$ of equation:

- $-k^{d+1} + (1+\varepsilon)(k+1)^d = 0$ if $\mathsf{EM} = \mathsf{PoA}_\varepsilon$,
- $-\frac{k^{d+1}}{d+1} + (1+\varepsilon)(k+1)^d = 0$ if $\mathsf{EM} = \mathsf{CR}^s_\varepsilon$,
- $-(2+\varepsilon)k^{d+1} + (1+\varepsilon)(k+1)^{d+1} = 0$ if $\mathsf{EM} = \mathsf{CR}^c_\varepsilon$.

A way to interpret the above tight bounds is provided by Remark 4.4. In particular, one can show that the exact value of $\gamma_{\mathsf{W}}(\mathsf{EM},\mathsf{W}(\mathcal{P}(d)))$ is attained by $\frac{kf(k)}{f(1)}$, where f is the monomial function defined as $f(t) = t^d$ for any $t \geq 0$, and k is the unique value satisfying constraint $\beta_{\mathsf{W}}(\mathsf{EM},k,1,f) \geq 0$ at equality.

Since the class of polynomial latency functions $\mathcal{P}(d)$ satisfies the hypothesis of Theorems 4.2 and 4.3, as a corollary we have that the values $\gamma_{\mathsf{W}}(\mathsf{EM},\mathsf{W}(\mathcal{P}(d)))$ considered above are tight bounds on the performance of load balancing instances, and even that of symmetric load balancing instances if the efficiency metric is the ε-price of anarchy; thus, for polynomial latency functions, the performance does not improve when assuming that the players' strategies are singletons.

See Figure 4.2 for some numerical comparisons.

d	$\mathsf{PoA}_0(\mathsf{W}(\mathcal{P}(d)))$ [7, 20]	$\mathsf{CR}_0^s(\mathsf{W}(\mathcal{P}(d)))$	$\mathsf{CR}_0^c(\mathsf{W}(\mathcal{P}(d)))$ [48]
1	2.618 [12, 54]	7.464 [63]	5.828 [54]
2	9.909	90.3	56.94
3	47.82	1,521	780.2
4	277	32,896	13,755
5	1,858	868,567	296,476
6	14,099	27,089,557	7,553,550
7	118,926	974,588,649	222,082,591
8	1,101,126	39,729,739,895	7,400,694,480
⋮	⋮	⋮	⋮
∞	$(\Theta(d/\log(d)))^{d+1}$	$(\Theta(d))^{d+1}$	$(\Theta(d))^{d+1}$

Fig. 4.2 The price of anarchy and the competitive ratio of exact one-round walks (generated by selfish or cooperative players) in weighted congestion games or load balancing games with polynomial latency functions of maximum degree d.

4.4 Application to Load Balancing Problems with Unrelated Machines

For a given increasing and continuous function $f : \mathbb{R}_{\geq 0} \to \mathbb{R}_{\geq 0}$ with $f(0) = 0$, one can use the same arguments of Theorem 4.1 to give upper bounds on the performance of load balancing games with unrelated machines, when the considered social function is the f-weighted total load WTL_f (for this class of games, refer to Section 2.6). Let $\gamma_{\mathsf{W}}(\mathsf{EM},x,k,o,f)$ be the quantity defined in (4.2).

Theorem 4.4. *The ε-approximate price of anarchy of any load balancing game* LB *with unrelated machines, with respect to the f-weighted total load, is at most*

$$\mathsf{PoA}_\varepsilon(\mathsf{LB}) \leq \inf_{x\geq 1} \sup_{k,o>0} \gamma_{\mathsf{W}}(\mathsf{PoA}_\varepsilon,x,k,o,f). \quad (4.22)$$

Proof (Sketch). Fix a load balancing game $\mathsf{LB} := (I,J,(w_{i,j})_{i,j})$ with unrelated machines, and let $\boldsymbol{\sigma}$ and $\boldsymbol{\sigma}^*$ be an equilibrium and an optimal assignment, respectively. Consider the following linear program $\mathsf{LP1}_f$ in variables $(\alpha_i)_{i\in I}$, in place of the linear program LP1 defined in Theorem 4.1:

$$\mathsf{LP1}_f: \quad \max \sum_{i\in I} \alpha_i w_i(\boldsymbol{\sigma}) f(w_i(\boldsymbol{\sigma})) \quad (4.23)$$

$$s.t. \quad \alpha_{\sigma_j} f(w_{\sigma_j}(\boldsymbol{\sigma})) \leq (1+\varepsilon)\alpha_{\sigma_j^*} f(w_{\sigma_j^*}(\boldsymbol{\sigma}) + w_j), \quad \forall j \in J \quad (4.24)$$

$$\sum_{i\in I} \alpha_i w_i(\boldsymbol{\sigma}^*) f(w_i(\boldsymbol{\sigma}^*)) = 1 \quad (4.25)$$

$$\alpha_i \geq 0, \quad \forall i \in I.$$

Observe that, by setting $\alpha_i = 1$ for any $i \in I$, (4.23) is the f-weighted total load of $\boldsymbol{\sigma}$, (4.24) is a relaxed ε-Nash equilibrium condition for any job $j \in J$, and the left-hand part of (4.25) is the f-weighted total load of $\boldsymbol{\sigma}^*$. Thus, by similar considerations as those in Theorem 4.1, we get that the objective function of $\mathsf{LP1}_f$ is an upper bound on the ε-price of anarchy of linear load balancing games with unrelated machines. Now, by proceeding as in the proof of Theorem 4.1, one can show that the right-hand part of (4.22) is an upper bound on the ε-price of anarchy of linear load balancing with unrelated machines. □

By using a similar approach as in the above theorem to construct the initial linear program, and by proceeding as in Theorem 4.1 (in particular, by resorting to the competitive ratio analysis) one can also prove the following theorem.

Theorem 4.5. *The competitive ratio of ε-approximate one-round walks involving selfish players of any load balancing game* LB *with unrelated machines, with respect to the f-weighted total load, is at most*

$$\mathsf{PoA}_\varepsilon(\mathsf{LB}) \leq \inf_{x \geq 1} \sup_{k,o>0} \gamma_{\mathsf{W}}(\mathsf{EM}, x, k, o, f), \tag{4.26}$$

where $\mathsf{EM} = \mathsf{CR}_\varepsilon^s$. *The above upper bound also holds for* $\mathsf{EM} = \mathsf{CR}_\varepsilon^c$ *(i.e., for cooperative players) if the function f is semi-convex.*

Remark 4.5. As a byproduct of our results, we also have that the upper bound of Theorem 4.5, for cooperative players and $\varepsilon = 0$, is an upper bound on the approximation guarantee of the classical greedy algorithm for (offline and online) load balancing problems under the f-weighted total load.

Remark 4.6. As the load balancing instances considered in Theorems 4.2 and 4.3 can be seen as particular cases of load balancing instances with related machines and restricted assignments (unrestricted assignments when the game is symmetric), we can easily turn the lower bounds of Section 4.2 into tight lower bounds for the efficiency guarantees shown in Theorems 4.4 and 4.5.

4.5 The Unweighted Case

By resorting to the same approach adopted for weighted games, in Theorem 4.6, we provide quasi-explicit formulas to compute upper bounds for unweighted games with general latency functions. Then, in Lemma 4.4, we provide some parameters that can be used to derive tight lower bounds. The proofs are omitted as they are equivalent to those of Theorem 4.1 and Lemma 4.1.

We first give some preliminary definitions. Given $x \geq 1$, $k \in \mathbb{Z}_{\geq 0}$, $o \in \mathbb{N}$, and a latency function f, let

$$\beta_{\mathsf{U}}(\mathsf{EM},k,o,f) := \begin{cases} -kf(k)+(1+\varepsilon)of(k+1) & \text{if } \mathsf{EM}=\mathsf{PoA}_\varepsilon, \\ -\sum_{h=1}^{k} f(h)+(1+\varepsilon)of(k+1) & \text{if } \mathsf{EM}=\mathsf{CR}^s_\varepsilon, \\ -kf(k)+(1+\varepsilon)((k+1)f(k+1)-kf(k)) & \text{if } \mathsf{EM}=\mathsf{CR}^c_\varepsilon. \end{cases}$$

Given $\mathsf{EM} \in \{\mathsf{PoA}_\varepsilon, \mathsf{CR}^s_\varepsilon, \mathsf{CR}^c_\varepsilon\}$, let

$$\gamma_{\mathsf{U}}(\mathsf{EM},x,k,o,f) := \frac{kf(k)+x\cdot\beta_{\mathsf{U}}(\mathsf{EM},k,o,f)}{of(o)};$$

furthermore, given a class of unweighted congestion games $\mathcal{G}$, let

$$\gamma_{\mathsf{U}}(\mathsf{EM},\mathcal{G}) := \inf_{x\geq 1} \sup_{f\in\mathcal{C}(\mathcal{G}),k\in\mathbb{Z}_{\geq 0},o\in\mathbb{N}} \gamma_{\mathsf{U}}(\mathsf{EM},x,k,o,f).$$

Theorem 4.6. *Let $\mathcal{G}$ be a class of unweighted congestion games. For any efficiency metric $\mathsf{EM} \in \{\mathsf{PoA}_\varepsilon, \mathsf{CR}^s_\varepsilon\}$, we have that $\mathsf{EM}(\mathcal{G}) \leq \gamma_{\mathsf{U}}(\mathsf{EM},\mathcal{G})$. This fact holds for $\mathsf{EM} = \mathsf{CR}^c_\varepsilon$ if the latency functions of $\mathcal{C}(\mathcal{G})$ are semi-convex.*

Lemma 4.4. *Let $\mathcal{G}$ be a class of unweighted congestion games. For each $M < \gamma_{\mathsf{U}}(\mathsf{EM},\mathcal{G})$, with $\mathsf{EM} \in \{\mathsf{PoA}_\varepsilon, \mathsf{CR}^s_\varepsilon\}$, one of the following cases holds:*

- **Case 1:** *there exists a latency function $f \in \mathcal{C}(\mathcal{G})$ and two integer numbers $k > 0, o > 0$ such that*
$$M < \frac{kf(k)}{of(o)}$$
and $\beta_{\mathsf{U}}(\mathsf{EM},k,o,f) \geq 0$.
- **Case 2:** *there exist two latency functions $f_1, f_2 \in \mathcal{C}(\mathcal{G})$ and four integer numbers $k_2 \geq 0$, $k_1, o_1, o_2 > 0$ such that*
$$M < \frac{\alpha_1 k_1 f_1(k_1)+\alpha_2 k_2 f_2(k_2)}{\alpha_1 o_1 f_1(o_1)+\alpha_2 o_2 f_2(o_2)},$$
where $\alpha_1 := \beta_{\mathsf{U}}(\mathsf{EM},k_2,o_2,f_2) > 0$ and $\alpha_2 := -\beta_{\mathsf{U}}(\mathsf{EM},k_1,o_1,f_1) > 0$.

This fact holds for $\mathsf{EM} = \mathsf{CR}^c_\varepsilon$ if the latency functions of $\mathcal{C}(\mathcal{G})$ are semi-convex.

By exploiting Lemma 4.4, as done for weighted games, we show that the upper bounds of Theorem 4.6 are tight even for load balancing games, under mild assumptions.

Theorem 4.7. *Let $\mathcal{C}$ be a class of latency functions that is closed under ordinate scaling. Then $\gamma_{\mathsf{U}}(\mathsf{PoA}_\varepsilon, \mathsf{U}(\mathcal{C})) = \mathsf{PoA}_\varepsilon(\mathsf{ULB}(\mathcal{C})) = \mathsf{PoA}_\varepsilon(\mathsf{U}(\mathcal{C}))$.*

Theorem 4.8. *Let $\mathcal{C}$ be a class of latency functions that is closed under ordinate scaling. Then $\gamma_{\mathsf{U}}(\mathsf{CR}^s_\varepsilon, \mathsf{U}(\mathcal{C})) = \mathsf{CR}^s_\varepsilon(\mathsf{U}(\mathcal{C})) = \mathsf{CR}^s_\varepsilon(\mathsf{ULB}(\mathcal{C}))$. If the functions of $\mathcal{C}$ are semi-convex, we have that $\gamma_{\mathsf{U}}(\mathsf{CR}^c_\varepsilon, \mathsf{U}(\mathcal{C})) = \mathsf{CR}^c_\varepsilon(\mathsf{U}(\mathcal{C})) = \mathsf{CR}^c_\varepsilon(\mathsf{ULB}(\mathcal{C}))$.*

To show the above theorems, we fix an arbitrary $M < \gamma_{\mathsf{U}}(\mathsf{EM},\mathcal{G})$, and by using a similar approach as in Theorems 4.2 and 4.3, we construct a class of load balancing games $\mathsf{LB}_s(k_1,k_2,o_1,o_2,f_1,f_2)$ parametrized by four non-negative integers k_1,k_2,o_1,o_2, two latency functions f_1,f_2 and an integer s. The above parameters are those provided in Lemma 4.4, where we set $k_1=k_2=k$, $o_1=o_2=o$, and $f_1=f_2=f$ if the first case of the lemma is verified. By using a similar proof as in Theorems 4.2 and 4.3, we show that $\mathsf{EM}(\mathsf{LB}_s(k_1,k_2,o_1,o_2,f_1,f_2)) > M$ for a sufficiently large s.

The lower bound constructions used in Theorems 4.7 and 4.8 are quite technical and are omitted. Here, we only show two weaker versions of Theorems 4.7 and 4.8, in which we provide tight lower bounds holding for general congestion games only.

Theorem 4.9. *We have that* $\gamma_{\mathsf{U}}(\mathsf{PoA}_\varepsilon,\mathsf{U}(\mathcal{C})) = \mathsf{PoA}_\varepsilon(\mathsf{U}(\mathcal{C}))$ *for any class* $\mathcal{C}$ *of latency functions.*[4]

Proof. We first assume that $\mathcal{C}$ is closed under ordinate-scaling.

Let $M < \gamma_{\mathsf{U}}(\mathsf{PoA}_\varepsilon,\mathsf{U}(\mathcal{C}))$, and let k_1,k_2,o_1,o_2,f_1,f_2 be the parameters provided by Lemma 4.4, where we set $k_1=k_2=k$, $o_1=o_2=o$, and $f_1=f_2=f$ if the first case of the lemma is verified. Let $\mathsf{CG} := \mathsf{CG}(k_1,k_2,o_1,o_2,f_1,f_2)$ be a congestion game defined as follows:

- we have a set N of $n := \max\{k_1+o_1,k_2+o_2\}$ players;
- the set of resources is $E := E_1 \cup E_2$, where $E_1 := \{e_{1,1},\dots e_{1,n}\}$ and $E_2 := \{e_{2,1},\dots e_{2,n}\}$;
- each player $i \in N$ has two strategies s_i, s_i^* only, namely,

$$\begin{aligned} s_i &:= \{e_{1,i},\dots,e_{1,i+k_1-1}\} \cup \{e_{2,i},\dots,e_{2,i+k_2-1}\} \text{ and} \\ s_i^* &:= \{e_{1,i+k_1},\dots,e_{1,i+k_1+o_1-1}\} \cup \{e_{2,i+k_2},\dots,e_{2,i+k_2+o_2-1}\}, \end{aligned}$$

  where the resource indexes are cyclical (i.e., $e_{j,i+n} := e_{j,i}$);
- for any $j \in [2]$, the latency function of each resource $e \in E_j$ is defined as $\ell_j(x) = \alpha_j \cdot f_j(x)$ for any $x \geq 0$, where $\alpha_1 := \beta_{\mathsf{U}}(\mathsf{EM},k_2,o_2,f_2) > 0$ and $\alpha_2 := -\beta_{\mathsf{U}}(\mathsf{EM},k_1,o_1,f_1) > 0$ if the second case of Lemma 4.4 holds, and $\alpha_1 := \alpha_2 := 1$ if the first case holds.

Let $\boldsymbol{\sigma}$ (resp. $\boldsymbol{\sigma}^*$) be the strategy profile in which each player i plays strategy s_i (resp. s_i^*). As

$$\begin{aligned} c_i(\boldsymbol{\sigma}) &= \alpha_1 k_1 f_1(k_1) + \alpha_2 k_2 f_2(k_2) \\ &= (1+\varepsilon)(\alpha_1 o_1 f_1(k_1+1) + \alpha_2 o_2 f_2(k_2+1)) \\ &= (1+\varepsilon) c_i(\boldsymbol{\sigma}_{-i}, s_i^*) \end{aligned}$$

[4] An equivalent version of Theorem 4.9 has appeared in Roughgarden [158], where the upper bound is written in terms of the smoothness framework and the lower bounding instance is the the same as the one provided here. We also point out that a similar instance has been previously introduced by Aland et al. [7] for the case of polynomial latency functions.

for any player $i \in N$, we have that $\boldsymbol{\sigma}$ is an ε-approximate pure Nash equilibrium. Furthermore, we have that

$$\mathsf{PoA}_{\varepsilon}(\mathsf{CG}) \geq \frac{\mathsf{SUM}(\boldsymbol{\sigma})}{\mathsf{SUM}(\boldsymbol{\sigma}^*)} = \frac{(\alpha_1 k_1 f_1(k_1) + \alpha_2 k_2 f_2(k_2))n}{(\alpha_1 o_1 f_1(o_1) + \alpha_2 o_2 f_2(o_2))n} > M,$$

where the last inequality follows from Lemma 4.4. Thus, CG is a tight lower bounding instance.

Now, if $\mathcal{C}$ is not ordinate-scaling, we consider again the lower bounding instance CG defined above. This game guarantees a tight lower bound, but its latencies might not belong to $\mathcal{C}$. Anyway, we can easily modify CG in such a way that each resource of group E_j (for $j \in [2]$) is replaced with a number of resources approximately proportional to α_j and having latency f_j, so that the new instance has latencies in $\mathcal{C}$, the equilibria conditions are guaranteed, and the resulting price of anarchy is again above M. □

Theorem 4.10. *We have that* $\gamma_{\mathsf{U}}(\mathsf{CR}^s_{\varepsilon}, \mathsf{U}(\mathcal{C})) = \mathsf{CR}^s_{\varepsilon}(\mathsf{U}(\mathcal{C}))$ *for any class* $\mathcal{C}$ *of latency functions. Furthermore, if the latency functions of* $\mathcal{C}$ *are semi-convex, we get* $\gamma_{\mathsf{U}}(\mathsf{CR}^c_{\varepsilon}, \mathsf{U}(\mathcal{C})) = \mathsf{CR}^c_{\varepsilon}(\mathsf{U}(\mathcal{C}))$.[5]

Proof. As in Theorem 4.9, we can assume without loss of generality that $\mathcal{C}$ is closed under ordinate-scaling. We first show the theorem for $\mathsf{EM} = \mathsf{CR}^s_{\varepsilon}$. Let $M < \gamma_{\mathsf{U}}(\mathsf{EM}, \mathsf{U}(\mathcal{C}))$, and let $k_1, k_2, o_1, o_2, f_1, f_2$ be the parameters provided by Lemma 4.4, where we set $k_1 = k_2 = k$, $o_1 = o_2 = o$, and $f_1 = f_2 = f$ if the first case of the lemma is verified. Let $\hat{n} := \max\{k_1 + o_1, k_2 + o_2\}$. For any integer $n > \hat{n}$, let $\mathsf{CG}_n := \mathsf{CG}_n(k_1, k_2, o_1, o_2, f_1, f_2)$ be a congestion game defined as follows:

- $N := [n]$ is the set of players,
- the set of resources is $E := E_1 \cup E_2$, where $E_1 := \{e_{1,1}, \ldots e_{1,n+\hat{n}}\}$ and $E_2 := \{e_{2,1}, \ldots e_{2,n+\hat{n}}\}$;
- each player $i > \hat{n}$ has two strategies s_i, s_i^* only, namely,

$$s_i := \{e_{1,i}, \ldots, e_{1,i+k_1-1}\} \cup \{e_{2,i}, \ldots, e_{2,i+k_2-1}\} \text{ and}$$
$$s_i^* := \{e_{1,i-o_1}, \ldots, e_{1,i-1}\} \cup \{e_{2,i-o_2}, \ldots, e_{2,i-1}\};$$

- each player $i \leq \hat{n}$ is constrained to play strategy s_i defined as in the above case;
- for any $j \in [2]$, the latency function of each resource $e \in E_j$ is defined as $\ell_j(x) = \alpha_j \cdot f_j(x)$ for any $x \geq 0$, where $\alpha_1 := \beta_{\mathsf{U}}(\mathsf{EM}, k_2, o_2, f_2) > 0$ and $\alpha_2 := -\beta_{\mathsf{U}}(\mathsf{EM}, k_1, o_1, f_1) > 0$ if the second case of Lemma 4.4 holds, and $\alpha_1 := \alpha_2 := 1$ if the first case holds.

Let $\boldsymbol{\sigma}_n$ (resp. $\boldsymbol{\sigma}_n^*$) be the strategy profile in which each player i plays strategy s_i (resp. s_i^* if $i > \hat{n}$, and s_i if $i \leq \hat{n}$). Let $\boldsymbol{\tau}_n := (\boldsymbol{\sigma}_n^1, \boldsymbol{\sigma}_n^2, \ldots, \boldsymbol{\sigma}_n^n)$ be the online process in which all the players enter the game in increasing order of their indexes, and

[5] A similar lower bounding instance as the one presented in Theorem 4.10 has been introduced by Bilò et al. [31] for the restricted case of affine functions.

choose strategy s_i; we observe that $\boldsymbol{\sigma}_n$ is the strategy profile obtained at the end (i.e., $\boldsymbol{\sigma}_n = \boldsymbol{\sigma}_n^n$). For any player $i > \hat{n}$, we have that

$$\begin{aligned} c_i(\boldsymbol{\sigma}_n^i) &= \alpha_1 \sum_{h=1}^{k_1} f_1(h) + \alpha_2 \sum_{h=1}^{k_2} f_2(h) \\ &= (1+\varepsilon)(\alpha_1 o_1 f_1(k_1+1) + \alpha_2 o_2 f_2(k_2+1)) \\ &= (1+\varepsilon) c_i(\boldsymbol{\sigma}_{n-1}^i, s_i^*), \end{aligned}$$

thus $\boldsymbol{\tau}_n$ is an ε-approximate one-round walk involving selfish players. Furthermore we have that

$$\begin{aligned} \mathsf{CR}_\varepsilon^s(\mathcal{C}) &= \limsup_{n\to\infty} \mathsf{CR}_\varepsilon^s(\mathsf{CG}_n) \\ &\geq \lim_{n\to\infty} \frac{\mathsf{SUM}(\boldsymbol{\sigma}_n)}{\mathsf{SUM}(\boldsymbol{\sigma}_n{}^*)} \\ &= \frac{(\alpha_1 k_1 f_1(k_1) + \alpha_2 k_2 f_2(k_2))n}{(\alpha_1 o_1 f_1(o_1) + \alpha_2 o_2 f_2(o_2))n} \\ &> M, \end{aligned}$$

where the last inequality follows from Lemma 4.4. Thus, by the arbitrariness of M, the claim follows.

Now, if $\mathsf{EM} = \mathsf{CR}_\varepsilon^c$, we can reuse the same lower bounding instance considered for the case of selfish players, where the parameters $k_1, k_2, o_1, o_2, f_1, f_2$ are again defined according to Lemma 4.4, but with respect to $\mathsf{EM} = \mathsf{CR}_\varepsilon^c$. Let us define $\boldsymbol{\sigma}_n$, $\boldsymbol{\sigma}_n^*$, $\boldsymbol{\tau}_n$ as in the above case. Similarly to the case of selfish players, one can show that $\boldsymbol{\tau}$ is an ε-approximate one-round walk involving cooperative players, and that $\lim_{n\to\infty} \frac{\mathsf{SUM}(\boldsymbol{\sigma}_n)}{\mathsf{SUM}(\boldsymbol{\sigma}_n{}^*)}$ approaches $\frac{(\alpha_1 k_1 f_1(k_1)+\alpha_2 k_2 f_2(k_2))n}{(\alpha_1 o_1 f_1(o_1)+\alpha_2 o_2 f_2(o_2))n}$ as n tends to infinite. Thus, the claim follows for cooperative players too. □

Remark 4.7. By exploiting the lemmas and the theorems of this section, we have that similar observations and characterizations as those provided in Remarks 4.2 and 4.4 can be easily reformulated for the unweighted case, too.

Application to Polynomial Latency Functions

Let $\gamma_{\mathsf{U}}(\mathsf{EM}, \mathsf{U}(\mathcal{P}(d)))$ with $\mathsf{EM} \in \{\mathsf{PoA}_\varepsilon, \mathsf{CR}_\varepsilon^s, \mathsf{CR}_\varepsilon^c\}$ be the quantity defined at the beginning of the section, where $\mathcal{P}(d)$ is the class of polynomial latency functions of maximum degree d. By Theorem 4.6, $\gamma_{\mathsf{U}}(\mathsf{EM}, \mathsf{U}(\mathcal{P}(d)))$ is an upper bound on the efficiency metrics $\mathsf{EM} \in \{\mathsf{PoA}_\varepsilon, \mathsf{CR}_\varepsilon^s, \mathsf{CR}_\varepsilon^c\}$ for unweighted congestion games and load balancing games with polynomial latency functions of maximum degree d.

$\gamma_{\mathsf{U}}(\mathsf{PoA}_\varepsilon, \mathsf{U}(\mathcal{P}(d)))$ has been evaluated in Aland et al. [7] and Christodoulou et al. [61]. $\gamma_{\mathsf{U}}(\mathsf{CR}_\varepsilon^c, \mathsf{U}(\mathcal{P}(d)))$ has been evaluated in Caragiannis et al. [54] for exact pure

Nash equilibria and affine functions. By using similar arguments as in Aland et al. [7], one can easily compute the exact value of $\gamma_{\mathsf{U}}(\mathsf{CR}^c_\varepsilon, \mathsf{U}(\mathcal{P}(d)))$ for any $\varepsilon > 0$.

The above values of $\mathsf{EM} \in \{\mathsf{PoA}_\varepsilon, \mathsf{CR}^c_\varepsilon\}$ can be also represented according to Remark 4.7. In particular, let k be the unique real solution of equation $\beta_{\mathsf{U}}(\mathsf{EM}, k, 1, f) = 0$, where f is the monomial function defined as $f(t) = t^d$. If k is not integer, we have that $\gamma_{\mathsf{U}}(\mathsf{EM}, \mathsf{U}(\mathcal{P}(d)))$ is equal to $\frac{\alpha_1 k_1 f_1(k_1) + \alpha_2 k_2 f_2(k_2)}{\alpha_1 o_1 f_1(o_1) + \alpha_2 o_2 f_2(o_2)}$, where $(k_1, k_2, o_1, o_2) := (\lfloor k \rfloor + 1, \lfloor k \rfloor, 1, 1)$, $f_1 := f_2 := f$, and α_1 and α_2 are defined as in Case 2 of Lemma 4.4 with respect to parameters $k_1, k_2, o_1, o_2, f_1, f_2$. Instead, if k is integer, we have that $\gamma_{\mathsf{U}}(\mathsf{EM}, \mathsf{U}(\mathcal{P}(d)))$ is equal to $\frac{kf(k)}{f(1)}$, i.e., Case 1 of Lemma 4.4 is satisfied by tuple $(k, 1, f)$.

Since the class of polynomial latency functions $\mathcal{P}(d)$ satisfies the hypothesis of Theorems 4.7 and 4.8, as a corollary we have that the values of $\gamma_{\mathsf{U}}(\mathsf{EM}, \mathsf{U}(\mathcal{P}(d)))$ considered above are tight upper bounds on the performance of load balancing games, thus matching the performance of general congestion games.

Regarding the ε-approximate one-round walks generated by selfish players, the exact value of $\gamma_{\mathsf{U}}(\mathsf{CR}^s_\varepsilon, \mathsf{U}(\mathcal{P}(d)))$ has been provided in Bilò [22] for $d \in [3]$ and $\varepsilon = 0$, thus, by Theorem 4.8, such a value is tight even for load balancing games. For more general polynomial latency functions with maximum degree $d > 3$ and for any $\varepsilon \geq 0$, an upper bound on $\gamma_{\mathsf{U}}(\mathsf{CR}^s_\varepsilon, \mathsf{U}(\mathcal{P}(d)))$ can be trivially obtained by reusing the upper bounds holding for the weighted case; for exact one-round walks, a better upper bound has been provided in Klimm et al. [110]. By Remark 4.7, we can get good lower bounds on the performance of ε-approximate one-round walks, having a similar representation as in the cases of $\mathsf{EM} \in \{\mathsf{PoA}_\varepsilon, \mathsf{CR}^c_\varepsilon\}$. In particular, let k be the highest non-negative integer such that $\beta_{\mathsf{U}}(\mathsf{CR}^s_\varepsilon, k, 1, f) \geq 0$, where f is the monomial function defined as $f(t) = t^d$. If $\beta_{\mathsf{U}}(\mathsf{CR}^s_\varepsilon, k, 1, f) > 0$ (resp. $\beta_{\mathsf{U}}(\mathsf{CR}^s_\varepsilon, k, 1, f) = 0$), we can consider the lower bounding instances used in Theorems 4.8 or 4.10, parametrized by $(k+1, k, 1, 1, f, f)$ (resp. $(k, k, 1, 1, f, f)$). By Remark 4.7, the performance of such instances is equal to $\frac{\alpha_1 k_1 f_1(k_1) + \alpha_2 k_2 f_2(k_2)}{\alpha_1 o_1 f_1(o_1) + \alpha_2 o_2 f_2(o_2)}$ (resp. $\frac{kf(k)}{of(o)}$), where α_1 and α_2 are defined as in Lemma 4.4.

See Figure 4.3 for some numerical comparisons.

4.6 Concluding Remarks and Related Work

Most of the results presented in this Chapter are based on the preliminary works of Bilò and Vinci [24, 25] (see [30] for the arXiv full-version), who show that the performance of load balancing games, under mild assumptions on the latency functions, is not better than that of general congestion games. The results of Bilò and Vinci presented here have addressed several open questions on the performance of load balancing games:

- Bhawalkar et al. [20] provided upper bounds on the price of anarchy of weighted congestion games and, under mild assumptions on the latency functions, such

d	$\mathsf{PoA}_0(\mathsf{U}(\mathcal{P}(d)))$ [7, 92]	$\mathsf{CR}_0^s(\mathsf{U}(\mathcal{P}(d)))$	$\mathsf{CR}_0^c(\mathsf{U}(\mathcal{P}(d)))$
1	2.5 [12, 54, 59]	4.236 [31, 63]	5.66 [54]
2	9.583	37.58 [22]	55.46
3	41.54	527.3 [22]	755.2
4	267.6	9,387 (L.B.)	13,170
5	1,514	201,401 (L.B.)	289,648
6	12,345	5,276,150 (L.B.)	7,174,495
7	98,734	151,192,413 (L.B.)	220,349,064
8	802,603	5,287,749,084 (L.B.)	7,022,463,077
⋮	⋮	⋮	⋮
∞	$(\Theta(d/\log(d)))^{d+1}$	$(\Theta(d))^{d+1}$	$(\Theta(d))^{d+1}$

Fig. 4.3 The competitive ratio of exact one-round walks generated by cooperative players in unweighted load balancing games with polynomial latency functions of maximum degree d. The acronym "L.B." means that the reported values are lower bounds only; for the upper bounds, the reader can refer to the values reported for weighted games (Figure 4.2).

bounds are tight. Theorem 4.2 shows that the above bounds, under the same assumptions, are tight even for load balancing games. Furthermore, if the latency functions are unbounded, the above tight bounds hold even for the subclass of symmetric load balancing games; this result generalizes some findings of Bhawalkar et al. [20], in which a similar limitation was shown for games with polynomial latencies only.

- The specific results on the concept of one-round walks have multiple implications, too. For instance, they close the gap between upper and lower bounds on the competitive ratio of exact one-round walks generated by selfish players, for congestion games with affine latency functions. Indeed, Christodoulou et al. [63] showed that $\gamma_{\mathsf{W}}(\mathsf{CR}_0^s, \mathsf{W}(\mathcal{P}(1))) = 2\sqrt{3}+4 \approx 7.464$ is an upper bound, and Bilò et al. [31] provided a lower bound of $2+\sqrt{5} \approx 4.236$ (holding even for unweighted congestion games); closing the above gap has been left as an open question. By exploiting Theorem 4.3, we get that the upper bound ≈ 7.464 provided in [63] is tight, even for load balancing games.
- Christodoulou et al. [63] and Klimm et al. [110] provided upper bounds for the competitive ratio of general congestion games with affine and polynomial latency functions, respectively, but tight lower bounds were missing. Theorem 4.3 automatically guarantees that such bounds are tight, even for load balancing games. If all functions in $\mathcal{C}$ are semi-convex, the same limitation applies to the competitive ratio of approximate one-round walks generated by cooperative players; this fact generalizes results from Awerbuch et al. [11], Caragiannis [49] and Caragiannis et al. [54], which hold only with respect to exact one-round walks for games with polynomial latency functions.
- Relatively to the unweighted case, the results obtained in Theorems 4.7 and 4.8 generalize some findings of Caragiannis et al. [54] and Gairing and Schoppmann [92], which hold only with respect to pure Nash equilibria and polyno-

mial latency functions, and generalize a result of Caragiannis et al. [54] and Suri et al. [170] which holds only with respect to exact one-round walks generated by cooperative players in games with affine latency functions.
- Bilò et al. [31] showed that the upper bound on the competitive ratio of exact one-round walks involving selfish players provided by Christodoulou et al. [63] for the case of affine latencies is tight. However, the lower bounding instance provided by Bilò et al. [31] is a general congestion game, and the authors left as an open problem the case of singleton strategies (for which the gap between upper and lower bound was not close yet). Theorem 4.8 can be directly used to construct a load balancing instance whose performance matches that of general congestion games, thus closing the above gap.

Other results obtained in Bilò and Vinci [24] are related to the quantification of the inefficiency in the restricted case of games with identical resources, and we give a brief overview of such results. In particular, we have that, if f is a non-decreasing, continuous, and semi-convex function, the approximate price of anarchy of weighted symmetric load balancing games with identical resources having latency function f is at most equal to

$$\sup_{x>0} \max_{\lambda\in(0,1)} \frac{\lambda x f(x) + (1-\lambda)[x]_{\varepsilon,f} f([x]_{\varepsilon,f})}{(\lambda x + (1-\lambda)[x]_{\varepsilon,f}) f(\lambda x + (1-\lambda)[x]_{\varepsilon,f})},$$

where $[x]_{\varepsilon,f} := \inf\{t \geq 0 : f(x) \leq (1+\varepsilon) f(x/2+t)\}$, and such upper bound is tight under some mild assumptions. This generalizes a result obtained by Gairing et al. [93], which holds only with respect to the price of anarchy under monomial latency functions. Finally, still for the case of identical resources, better lower bounds on the performance of exact one-round walks in load balancing games with unweighted players have been provided (this improves and generalizes a result of Caragiannis et al. [54] which holds only for affine latency functions).

The machinery defined in Bilò and Vinci [24] has been further explored to provide load balancing instances as tight lower bounds for several classes of congestion games. In this direction, the works of Benita et al. [18], Bilò et al. [35], Bilò et al. [39], Caragiannis et al. [56][6] have provided tight (or almost) tight on the performance of some variants of congestion games, and some of the tight lower bounds have been constructed by resorting to the modus-operandi described in this chapter. In particular:

- Benita et al. [18] study some variants of congestion games, called *θ-free flow games*; for this class of games, several upper and lower bounds have been constructed by resorting to the primal-dual method and by using similar load-balancing structures as those provided in Theorems 4.2 and 4.7.
- Bilò et al. [39], Bilò et al. [41] study the efficiency of weighted and unweighted congestion games under the *Nash social welfare*, that is a well-known social

[6] The works of Bilò et al. [35] and Benita et al. [18] are also discussed in the "Concluding Remarks and Related work" sections of Chapters 5 and 6, as they are inherent with the results presented in these chapters.

function defined as a weighted geometric mean of the players' costs. Some of the upper and lower bounds have been derived by reconsidering the approaches presented in this chapter.

- Bilò et al. [35] introduce a new notion of mixed equilibrium and study the efficiency of both weighted and unweighted affine congestion games under this framework. The lower bounds for the weighted (resp. unweighted) case have been derived using a construction similar to those adopted in Theorem 4.2 (resp. Theorem 4.7).
- Caragiannis et al. [56] analyze the efficiency of certain (scheduling) policies for machine-scheduling games/problems, that randomly schedule the jobs assigned to any machine, according to certain probability distributions. Almost tight lower bounds on the efficiency have been provided by resorting to similar load balancing constructions as in Theorem 4.2.

Chapter 5
Uniform Mixed Equilibria

In this chapter, we introduce the notion of uniform mixed equilibrium in atomic congestion games. Given an integer $\rho \geq 1$, a *ρ-uniform mixed strategy* is a mixed strategy in which exactly ρ pairwise disjoint (i.e., not sharing any resource) subsets of resources are chosen with uniform probabilities. Hence, a *ρ-uniform mixed equilibrium* is a ρ-uniform mixed profile, i.e., a tuple of ρ-uniform mixed strategies, one for each player, in which no player can lower her cost by deviating to another ρ-uniform mixed strategy. Observe that a 1-uniform mixed equilibrium coincides with a pure Nash equilibrium, but when $\rho > 1$, there are no correlations between the notion of ρ-uniform mixed equilibria and that of pure or mixed Nash equilibria of a given game. More precisely, for $\rho > 1$, ρ-uniform mixed strategies can be interpreted as an hybridization between the notions of pure and mixed strategies. In fact, although the cost incurred by a player needs to be evaluated in expectation (as it happens when adopting mixed strategies), the fact that probabilities are superimposed by the model limits the players' choices to deciding which strategies to play (as it happens when adopting pure strategies).

We apply this behavioral model to the class of congestion games with unweighted players. We first show that ρ-uniform mixed equilibria do exist, for each $\rho \geq 1$. This generalizes Rosenthal's Theorem [154], which applies to the basic case of $\rho = 1$. Then, for each $\rho \geq 1$, we derive tight bounds on the price of anarchy and the price of stability of ρ-uniform mixed equilibria in affine games; the bounds are obtained by resorting to the primal-dual method. For a numerical comparison, see the values reported in Table 5.1, where many lower bounds hold even under additional restrictions. The obtained results show that, as ρ increases, the price of anarchy and the price of stability of ρ-uniform mixed equilibria approach the value $4/3$, that is, the price of anarchy of non-atomic games [159]. This is in accordance with the intuition that, by arbitrarily splitting an atomic request over disjoint strategies, atomic congestion games tend to their non-atomic counterparts. The striking evidence of these findings, however, is that this happens for $\rho = 4$ already.

The equilibrium existence, as well as the PoA and PoS bounds, are obtained by exploiting the fact that, for each $\rho \geq 1$, any unweighted congestion game in which players adopt ρ-uniform mixed strategies is isomorphic to an unweighted

V. Bilò, C. Vinci, *Coping with Selfishness in Congestion Games*, Monographs in Theoretical Computer Science. An EATCS Series, https://doi.org/10.1007/978-3-031-30261-9_5

ρ	unweighted games		weighted games
	price of stability	price of anarchy	price of anarchy
1	$1+1/\sqrt{3}$	$5/2^{*}$	$(3+\sqrt{5})/2^{\dagger}$
2	$1+1/\sqrt{5}$	$5/3^{*}$	$2^{\dagger}$
3	$1+1/\sqrt{7}$	$(2\sqrt{7}-1)/3^{*}$	$(7+\sqrt{13})/6^{\dagger}$
4	$4/3^{\dagger}$	$4/3^{\dagger}$	$(9+\sqrt{17})/8^{\dagger}$
≥ 5	$4/3^{\dagger}$	$4/3^{\dagger}$	$4\rho^2/(3\rho^2-2\rho-1)^{\dagger}$

Table 5.1 The price of anarchy and the price of stability of ρ-uniform mixed equilibria in affine unweighted congestion games and the price of anarchy of ρ-uniform mixed equilibria in affine weighted congestion games, for each value of $\rho \geq 1$. Bounds labeled as $*$ extends to load balancing games, while bounds labeled as $\dagger$ applies to even symmetric load balancing games. For $\rho = 1$, the price of anarchy (resp. the price of stability) has been already determined in [12, 54, 59] (resp. [54, 60]).

congestion game in which players adopt pure strategies and whose latency functions are slightly different. We also provide the main intuitions on how, by using this isomorphism and adapting the techniques presented in Chapter 4, results for the price of anarchy can be generalized to general latency functions.

This chapter is based on the results obtained by Bilò et al. [35], and the interested reader can refer to its full version for the missing proofs and other results. In the "Concluding Remarks and Related Work" section at the end of the chapter, we also give a brief overview of the results obtained in [35] for the price of anarchy of weighted games (some values are listed in Table 5.1), and we point out several relevant references connected with the results presented here.

5.1 Uniform Mixed Equilibria and Fault-Tolerant Systems

The notion of uniform mixed equilibria can be used as a reasonable model for *risk-averse behavior* in resource selections games with failures (Penn et al. [145, 148, 149]). Consider, for instance, the case in which players need to send messages from a source to a destination in a communication network whose links are subject to adversarial failures. In particular, the adversary, upon knowledge of the strategies adopted by the players, has the power of corrupting any subset of k links, for a certain $k \geq 1$. Consider risk-averse players who are interested in choosing the fastest route among the ones minimizing the probability of failures (or keeping it below a certain threshold). In order to provide the adversary with the fewest possible information, each player will end up adopting a ρ-uniform mixed strategy, for some $\rho > k$. Indeed, this line of reasoning can be applied in other contexts as well, e.g., that of minimizing the probability that a message gets intercepted by an attacker.

Moreover, consider the special case of symmetric network congestion games, i.e., network congestion games in which all players share the same set of strategies.

This symmetric setting has been widely studied with respect to the analysis of efficiency of Nash equilibria (Fotakis et al. [90]) and to the (hardness of) computation of equilibria (Deeparnab et al. [82], Fabrikant et al. [83]). In a highly unsafe scenario (in which, for instance, rescue robots have to be used), the players' priority is often that of minimizing the probability of a failure, and then minimization of the expected cost comes only as a secondary objective. Thus, every player will adopt a ρ-uniform mixed strategy with $\rho = k$, where k is the maximum number of resource-disjoint strategies (i.e., paths) available to players. It is worth noticing that, in the context of network symmetric congestion games, it is well known that the value of k can be easily computed by a reduction to the max-flow problem (see, for instance, Ahuja et al. [6] for further details).

5.2 Model and Definitions

Congestion Model

Some of the notation provided in this subsection is equivalent or similar to that provided in Chapter 2. Furthermore, despite this chapter is essentially based on the sub-class of unweighted games, we will give more general definitions holding for weighted ones, too.

A *(weighted) congestion model* is defined by a tuple

$$\mathsf{CM} = (N, E, (\ell_e)_{e \in E}, (w_i)_{i \in N}, (\Sigma_i)_{i \in N}),$$

where N is a set of $n \geq 2$ players, E is a set of resources, $\ell_e : \mathbb{R}_{\geq 0} \to \mathbb{R}_{\geq 0}$ is the latency function of resource $e \in E$, and, for each $i \in N$, $w_i \geq 0$ is the weight of player i and $\Sigma_i \subseteq 2^E \setminus \emptyset$ is her set of strategies. An *unweighted congestion model* is a weighted congestion model such that $w_i = 1$ for each $i \in N$; in such case, we omit the set of weighs from the tuple modeling CM. Other classes of congestion models (e.g., load balancing congestion models, symmetric congestion models etc.) can be defined as in Chapter 2.

A *strategy profile* is an n-tuple of strategies $\boldsymbol{s} = (s_1, s_2, \ldots, s_n)$, that is a state in which each player $i \in N$ adopts pure strategy $s_i \in \Sigma_i$. When players adopt pure strategies, CM induces a congestion game CG(CM) (usually, when CM is clear from the context, we shall drop it from the notation). For a strategy profile $\boldsymbol{s}$, the *congestion* of resource $e \in E$ in $\boldsymbol{s}$, denoted as $k_e(\boldsymbol{s}) := \sum_{i \in N : e \in s_i} w_i$, is the total weight of the players using resource e in $\boldsymbol{s}$, (observe that, for unweighted games, $k_e(\boldsymbol{s})$ coincides with the number of users selecting resource e in $\boldsymbol{s}$). The cost of player i in $\boldsymbol{s}$ is defined as $cost_i^{\mathsf{CG}}(\boldsymbol{s}) = \sum_{e \in s_i} \ell_e(k_e(\boldsymbol{s}))$ (usually, when CG is clear from the context, we shall drop it from the notation). The quality of a strategy profile in CG(CM) is measured by using the *social function*

$$\mathsf{SUM}(\boldsymbol{s}) = \sum_{i \in N} w_i cost_i(\boldsymbol{s}) = \sum_{e \in E} k_e(\boldsymbol{s}) \ell_e(k_e(\boldsymbol{s})),$$

that is, the weighted sum of the players' costs. A *pure Nash equilibrium* for $\mathsf{CG}(\mathsf{CM})$ is a strategy profile $\boldsymbol{s}$ such that, for any player $i \in N$ and strategy $s_i' \in \Sigma_i$, $cost_i(\boldsymbol{s}) \leq cost_i(\boldsymbol{s}_{-i}, s_i')$. We denote by $\mathsf{Eq}(\mathsf{CG}(\mathsf{CM}))$ the set of pure Nash equilibria of a weighted congestion game $\mathsf{CG}(\mathsf{CM})$. The *price of anarchy* of a weighted congestion game $\mathsf{CG}(\mathsf{CM})$ is defined as

$$\mathsf{PoA}(\mathsf{CG}(\mathsf{CM})) = \max_{\boldsymbol{s} \in \mathsf{Eq}(\mathsf{CG}(\mathsf{CM}))} \frac{\mathsf{SUM}(\boldsymbol{s})}{\mathsf{SUM}(\boldsymbol{s}^*)},$$

where $\boldsymbol{s}^*$ is a *social optimum* for $\mathsf{CG}(\mathsf{CM})$. Similarly, the *price of stability* is defined as

$$\mathsf{PoS}(\mathsf{CG}(\mathsf{CM})) = \min_{\boldsymbol{s} \in \mathsf{Eq}(\mathsf{CG}(\mathsf{CM}))} \frac{\mathsf{SUM}(\boldsymbol{s})}{\mathsf{SUM}(\boldsymbol{s}^*)}.$$

ρ-Uniform Mixed Strategies

A mixed strategy for player i is a probability distribution σ_i defined over Σ_i, so that $\sigma_i(s)$ is the probability that player i plays strategy $s \in \Sigma_i$. With a little abuse of notation, we may also use σ_i to denote the random variable equal to a strategy s with probability $\sigma_i(s)$. We denote by $\mathsf{supp}(\sigma_i) = \{s \in \Sigma_i : \sigma_i(s) > 0\}$ the *support* of σ_i, that is, the set of strategies played with positive probability in σ_i. A *mixed profile* $\boldsymbol{\sigma}$ is an n-tuple of mixed strategies, i.e., $\boldsymbol{\sigma} = (\sigma_1, \sigma_2, \ldots, \sigma_n)$. Informally, $\boldsymbol{\sigma}$ is a state in which each player $i \in N$ picks her strategy according to probability distribution σ_i, independently from the choices of the other players. If σ_i is such that a pure strategy s_i is picked with probability 1 by player i, we write s_i instead of σ_i. The following definition provides the notion of ρ-uniform strategy, and the strategic behavior of the games considered here is based on this concept.

Definition 5.1 (ρ-uniform mixed strategy). Given an integer $\rho \geq 1$ and a weighted congestion model CM in which for each player $i \in N$ there exist at least ρ pairwise disjoint strategies in Σ_i, a *ρ-uniform mixed strategy* for player i is a mixed strategy σ_i such that $|\mathsf{supp}(\sigma_i)| = \rho$, $s_1 \cap s_2 = \emptyset$ for any $s_1, s_2 \in \mathsf{supp}(\sigma_i)$ with $s_1 \neq s_2$, and $\sigma_i(s) = 1/\rho$ for each $s \in \mathsf{supp}(\sigma_i)$, i.e., a mixed strategy in which player i plays exactly ρ pairwise disjoint strategies with uniform probability.

Denote by $\Delta_i^\rho(\mathsf{CM})$ the set of ρ-uniform mixed strategies for player i. A *ρ-uniform mixed profile* $\boldsymbol{\sigma} = (\sigma_1, \sigma_2, \ldots, \sigma_n)$ is an n-tuple of ρ-uniform mixed strategies, one for each player. When players adopt *ρ-uniform mixed strategies*, CM induces a ρ-uniform congestion game ρ-$\mathsf{CG}(\mathsf{CM})$ (again, when CM is clear from the context, we shall drop it from the notation). For a ρ-uniform mixed profile $\boldsymbol{\sigma}$, the *expected congestion* of resource $e \in E$ in $\boldsymbol{\sigma}$, denoted as $k_e(\boldsymbol{\sigma}) := \mathbb{E}_{\boldsymbol{s} \sim \boldsymbol{\sigma}}\left[\sum_{i \in N: e \in s_i} w_i\right]$, is the expected total weight of the players using resource e in $\boldsymbol{\sigma}$. The cost of player i in $\boldsymbol{\sigma}$

is defined as $cost_i^{\rho\text{-CG}}(\boldsymbol{\sigma}) = \mathbb{E}_{\boldsymbol{s}\sim\boldsymbol{\sigma}}\left[\sum_{e\in s_i} \ell_e(k_e(\boldsymbol{s}))\right]$ (again, when ρ-CG is clear from the context, we shall drop it from the notation). The quality of a ρ-uniform mixed profile in ρ-CG(CM) becomes

$$\mathsf{SUM}(\boldsymbol{\sigma}) = \mathbb{E}_{\boldsymbol{s}\sim\boldsymbol{\sigma}}\left[\sum_{i\in N} w_i cost_i(\boldsymbol{s})\right] = \sum_{e\in E} \mathbb{E}_{\boldsymbol{s}\sim\boldsymbol{\sigma}}\left[k_e(\boldsymbol{s})\ell_e(k_e(\boldsymbol{s}))\right],$$

that is, the expected weighted sum of the players' costs. A *ρ-uniform mixed equilibrium* for ρ-CG(CM) is a ρ-uniform mixed profile $\boldsymbol{\sigma}$ such that, for any player $i \in N$ and ρ-uniform mixed strategy $\sigma_i' \in \Delta_i^\rho(\mathsf{CM})$, $cost_i(\boldsymbol{\sigma}) \leq cost_i(\boldsymbol{\sigma}_{-i}, \sigma_i')$, that is, no player can lower her cost by unilaterally deviating to another ρ-uniform mixed strategy. We denote by $\mathsf{Eq}(\rho\text{-CG}(\mathsf{CM}))$ the set of ρ-uniform mixed equilibria of a weighted congestion game ρ-CG(CM). The price of anarchy of a ρ-uniform weighted congestion game ρ-CG(CM) is defined as

$$\mathsf{PoA}_\rho(\rho\text{-CG}(\mathsf{CM})) = \max_{\boldsymbol{\sigma}\in\mathsf{Eq}(\rho\text{-CG}(\mathsf{CM}))} \frac{\mathsf{SUM}(\boldsymbol{\sigma})}{\mathsf{SUM}(\boldsymbol{\sigma}^*)},$$

where $\boldsymbol{\sigma}^*$ is a *ρ-uniform social optimum* for ρ-CG(CM), that is a ρ-uniform mixed profile minimizing the social function. Similarly, the price of stability becomes

$$\mathsf{PoS}_\rho(\rho\text{-CG}(\mathsf{CM})) = \min_{\boldsymbol{\sigma}\in\mathsf{Eq}(\rho\text{-CG}(\mathsf{CM})} \frac{\mathsf{SUM}(\boldsymbol{\sigma})}{\mathsf{SUM}(\boldsymbol{\sigma}^*)}.$$

Given a ρ-uniform mixed strategy σ_i, let $E(\sigma_i) = \bigcup_{s\in\mathsf{supp}(\sigma_i)} s$ denote the set of resources contained by all strategies belonging to $\mathsf{supp}(\sigma_i)$[1]. For a ρ-uniform mixed profile $\boldsymbol{\sigma}$, the *ρ-maximum congestion* of resource e in $\boldsymbol{\sigma}$, denoted as $k_{\rho,e}(\boldsymbol{\sigma}) = \sum_{i\in N: e\in E(\sigma_i)} w_i$, is the congestion of e obtained if each agent i assigning non-null probability to a strategy s_i containing e decided to select s with probability 1 (that is, if s_i was the strategy chosen by i in a pure strategy profile).

5.3 Equilibrium Existence in ρ-Uniform Unweighted Games

In this section, we consider the case of ρ-uniform congestion games induced by unweighted congestion models. First, we show that uniform mixed equilibria are always guaranteed to exist for any class of latency functions. We achieve this result by defining a transformation from an unweighted congestion model to another, which induces an isomorphism between the ρ-uniform congestion game played on the former and the congestion game played on the latter. A fundamental role in this transformation is played by the following mapping among latency functions.

[1] Given $e \in E(\sigma_i)$, there exists a unique strategy of σ_i containing e, since strategies selected with non-null probability by each player are pairwise disjoint.

Definition 5.2. Given a positive integer ρ and a latency function ℓ, the *latency function derived from ℓ according to ρ* is the function

$$\lambda_{\ell,\rho}(x) := \frac{1}{\rho}\mathbb{E}_{j\sim\mathcal{B}\left(x-1,\frac{1}{\rho}\right)}[\ell(j+1)] = \sum_{j=0}^{x-1}\binom{x-1}{j}\left(\frac{1}{\rho}\right)^{j+1}\left(\frac{\rho-1}{\rho}\right)^{x-1-j}\ell(j+1),$$

where $\mathcal{B}(k,p)$ denotes the binomial distribution with parameters $k \in \mathbb{Z}_{\geq 0}$ and $p \in [0,1]$. A function ℓ' is *derived from ℓ according to ρ* if $\ell' = \lambda_{\ell,\rho}$.

Before presenting the promised transformation, we illustrate an example of the mapping $\lambda_{\ell,\rho}$ instantiated to a general affine latency function.

Lemma 5.1. *The latency function derived from $\ell(x) = \alpha x + \beta$ according to ρ is $\lambda_{\ell,\rho}(x) = \frac{\alpha(x+\rho-1)}{\rho^2} + \frac{\beta}{\rho}$.*

Proof. Given $x > 0$, we get

$$\begin{aligned}
\lambda_{\ell,\rho}(x) &= \frac{1}{\rho}\mathbb{E}_{j\sim\mathcal{B}\left(x-1,\frac{1}{\rho}\right)}[\alpha(j+1)+\beta] \\
&= \frac{\alpha}{\rho}\left(\mathbb{E}_{j\sim\mathcal{B}\left(x-1,\frac{1}{\rho}\right)}[j] + \mathbb{E}_{j\sim\mathcal{B}\left(x-1,\frac{1}{\rho}\right)}[1]\right) + \frac{1}{\rho}\mathbb{E}_{j\sim\mathcal{B}\left(x-1,\frac{1}{\rho}\right)}[\beta] \\
&= \frac{\alpha}{\rho}\left((x-1)\frac{1}{\rho}+1\right) + \frac{\beta}{\rho} \qquad (5.1) \\
&= \frac{\alpha(x+\rho-1)}{\rho^2} + \frac{\beta}{\rho},
\end{aligned}$$

where (5.1) holds since $\mathbb{E}_{j\sim\mathcal{B}(k,p)}[j] = kp$ for any $k \geq 0$ and $p \in [0,1]$. □

We are now ready to present our transformation. Let Λ be the function mapping CM and ρ to another congestion model $\Lambda(\mathsf{CM},\rho)$ defined as follows. For any integer $\rho \geq 1$, and given $\mathsf{CM} = (N,E,(\ell_e)_{e\in E},(\Sigma_i)_{i\in N})$ as input unweighted congestion model, let

$$\Lambda(\mathsf{CM},\rho) = \mathsf{CM}' = \left(N',E',(\ell'_e)_{e\in E},(\Sigma'_i)_{i\in N}\right)$$

denote the unweighted congestion model obtained from CM, such that $N' = N$, $E' = E$, $\Sigma'_i = \{E(\sigma_i) : \sigma_i$ is a ρ-uniform mixed strategy for player i in ρ-$\mathsf{CG}(\mathsf{CM})\}$ for each $i \in N$, and $\ell'_e = \lambda_{\ell_e,\rho}$ for each $e \in E'$. Moreover, for a ρ-uniform mixed profile $\boldsymbol{\sigma}$ for ρ-$\mathsf{CG}(\mathsf{CM})$, define $\boldsymbol{s}_\rho(\boldsymbol{\sigma})$ as the strategy profile for $\mathsf{CG}(\Lambda(\mathsf{CM},\rho))$ such that

$$\boldsymbol{s}_\rho(\boldsymbol{\sigma}) := (E(\sigma_1),E(\sigma_2),\ldots,E(\sigma_n)),$$

i.e., each player i plays in $\mathsf{CG}(\Lambda(\mathsf{CM},\rho))$ all the resources selected with positive probability when playing strategy σ_i in game ρ-$\mathsf{CG}(\mathsf{CM})$.

By using map Λ, we show that any ρ-uniform unweighted congestion game can be transformed into an equivalent unweighted congestion game.

Theorem 5.1. *Fix an integer $\rho \geq 1$ and an unweighted congestion model* CM. *We have that*

$$cost_i^{\rho\text{-}\mathsf{CG(CM)}}(\boldsymbol{\sigma}) = cost_i^{\mathsf{CG}(\Lambda(\mathsf{CM},\rho))}(\mathbf{s}_\rho(\boldsymbol{\sigma}))$$

for each ρ-uniform mixed profile $\boldsymbol{\sigma}$ of ρ-$\mathsf{CG(CM)}$, and for any $i \in N$.

Proof. Fix an integer $\rho \geq 1$ and an unweighted congestion model CM. Given a ρ-uniform mixed profile $\boldsymbol{\sigma}$ for ρ-$\mathsf{CG(CM)}$ and a player $i \in N$, we have

$$\begin{aligned}
& cost_i^{\rho\text{-}\mathsf{CG}}(\boldsymbol{\sigma}) \\
&= \sum_{s \in \mathsf{supp}(\sigma_i)} \overbrace{\mathbb{P}[\sigma_i(s) = s]}^{1/\rho} \sum_{e \in s} \sum_{j=0}^{k_{\rho,e}(\boldsymbol{\sigma})-1} \mathbb{P}[k_e(\boldsymbol{\sigma}_{-i}, s) = j+1]\ell_e(j+1) \\
&= \sum_{e \in E(\sigma_i)} \frac{1}{\rho} \sum_{j=0}^{k_{\rho,e}(\boldsymbol{\sigma})-1} \mathbb{P}[k_e(\boldsymbol{\sigma}_{-i}, E(\sigma_i)) = j+1]\ell_e(j+1) \\
&= \sum_{e \in E(\sigma_i)} \frac{1}{\rho} \sum_{j=0}^{k_{\rho,e}(\boldsymbol{\sigma})-1} \binom{k_{\rho,e}(\boldsymbol{\sigma})-1}{j} \left(\frac{1}{\rho}\right)^j \left(\frac{\rho-1}{\rho}\right)^{k_{\rho,e}(\boldsymbol{\sigma})-1-j} \ell_e(j+1) \\
&= \sum_{e \in E(\sigma_i)} \frac{1}{\rho} \mathbb{E}_{j \sim \mathcal{B}\left(k_{\rho,e}(\boldsymbol{\sigma})-1, \frac{1}{\rho}\right)}[\ell_e(j+1)] \\
&= \sum_{e \in E(\sigma_i)} \lambda_{\rho,\ell_e}(k_{\rho,e}(\boldsymbol{\sigma})) \\
&= cost_i^{\mathsf{CG}(\Lambda(\mathsf{CM},\rho))}(\mathbf{s}_\rho(\boldsymbol{\sigma})).
\end{aligned}$$

□

As a consequence, we obtain the existence of uniform mixed equilibria for each uniform congestion game induced by an unweighted congestion model, regardless of which are its latency functions. In particular, we extend Rosenthal's Theorem [154], by showing that, for each $\rho \geq 1$, any ρ-uniform unweighted congestion game admits an exact potential.

Theorem 5.2. *For each $\rho \geq 1$ and unweighted congestion model* CM, ρ-$\mathsf{CG(CM)}$ *admits an exact potential.*

Proof. Fix an integer $\rho \geq 1$ and an unweighted congestion model CM. By Rosenthal's Theorem [154], the function Φ defined as

$$\Phi(\boldsymbol{s}) := \sum_{e \in E} \sum_{j=1}^{k_e(\boldsymbol{s})} \lambda_{\ell_e,\rho}(k_e(\boldsymbol{s}))$$

is an exact potential function for $\mathsf{CG}(\Lambda(\mathsf{CM},\rho))$, i.e.,

$$cost_i^{\mathsf{CG}(\Lambda(\mathsf{CM},\rho))}(\mathbf{s}) - cost_i^{\mathsf{CG}(\Lambda(\mathsf{CM},\rho))}(\mathbf{s}_{-i}, \widehat{s_i}) = \Phi(\mathbf{s}) - \Phi(\mathbf{s}_{-i}, \widehat{s_i})$$

for any (pure) strategy profile $\boldsymbol{s}$ of $\mathsf{CG}(\Lambda(\mathsf{CM},\rho))$ and (pure) strategy $\widehat{s_i}$ of player i in $\mathsf{CG}(\Lambda(\mathsf{CM},\rho))$. By exploiting the isomorphism (via map Λ) shown in Theorem 5.1, one can easily show that the function $\Phi \circ \boldsymbol{s}_\rho$ is an exact potential function for ρ-CG(CM) (i.e., $cost_i^{\rho\text{-CG}}(\boldsymbol{\sigma}) - cost_i^{\rho\text{-CG}}(\boldsymbol{\sigma}_{-i}, \widehat{\sigma}_i) = \Phi(\boldsymbol{s}_\rho(\boldsymbol{\sigma})) - \Phi(\boldsymbol{s}_\rho(\boldsymbol{\sigma}_{-i}, \widehat{\sigma}_i))$ for any ρ-uniform strategy profile $\boldsymbol{\sigma}$ of ρ-CG and any ρ-uniform strategy $\widehat{\sigma}_i$ of player i in ρ-CG), and this concludes the proof. □

5.4 Performance of ρ-Uniform Unweighted Games

In this section we provide tight bounds on the price of anarchy and the price of stability of ρ-uniform unweighted congestion games, and we focus on the particular case of affine latencies.

The Price of Anarchy for Affine Latencies

In this subsection, we derive exact bounds on the price of anarchy of ρ-uniform congestion games induced by affine unweighted congestion models.

Theorem 5.3. *Fix an affine unweighted congestion model* CM. *For any* $\rho \geq 1$, *we have*

$$\mathsf{PoA}_\rho(\rho\text{-}\mathsf{CG}(\mathsf{CM})) \leq \begin{cases} \frac{5}{\rho+1} & \text{if } \rho \leq 2, \\ \frac{2\sqrt{7}-1}{3} & \text{if } \rho = 3, \\ \frac{4}{3} & \text{if } \rho \geq 4. \end{cases}$$

Proof. Given an integer $\rho \geq 1$, let CM be an arbitrary affine unweighted congestion model. Similarly as in classical unweighted games, we can assume without loss of generality that the latency functions are of type $\ell_e(x) = \alpha_e x$ (see [22]). As $\mathsf{PoA}_\rho(\rho\text{-}\mathsf{CG}(\mathsf{CM})) = \mathsf{PoA}(\mathsf{CG}(\Lambda(\mathsf{CM},\rho)))$ (by Theorem 5.1), it is sufficient showing that the desired upper bound holds for $\mathsf{PoA}(\mathsf{CG}(\Lambda(\mathsf{CM},\rho)))$.

Let $\boldsymbol{s} = (s_1, \ldots, s_n)$ and $\boldsymbol{s}^* = (s_1^*, \ldots, s_n^*)$ be a pure Nash equilibrium and a social optimum for $\mathsf{CG}(\Lambda(\mathsf{CM},\rho))$, respectively. By exploiting Lemma 5.1, we have that, for each $i \in N$, the inequality $cost_i(\boldsymbol{s}) \leq cost_i(\boldsymbol{s}_{-i}, s_i^*)$ becomes

$$\begin{aligned} 0 &\geq \sum_{e \in s_i} \lambda_{\ell_e,\rho}(k_e(\boldsymbol{s})) - \sum_{e \in s_i^*} \lambda_{\ell_e,\rho}(k_e(\boldsymbol{s}) + 1) \\ &\geq \sum_{e \in s_i} \alpha_e \left(\frac{k_e(\boldsymbol{s}) + \rho - 1}{\rho^2} \right) - \sum_{e \in s_i^*} \alpha_e \left(\frac{k_e(\boldsymbol{s}) + \rho}{\rho^2} \right). \end{aligned} \tag{5.2}$$

By applying the primal-dual method as in Example 3.1 and in Theorem 4.1, and by exploiting Lemma 5.1 to explicate $\mathsf{SUM}(\boldsymbol{s})$ and $\mathsf{SUM}(\boldsymbol{s}^*)$, we get the following linear program in variables α_es, whose optimal value is an upper bound on the price of anarchy:

$$\text{LP}: \max \quad \overbrace{\sum_{e\in E}\alpha_e k_e(\boldsymbol{s})\left(\frac{k_e(\boldsymbol{s})+\rho-1}{\rho^2}\right)}^{\text{SUM}(\boldsymbol{s})}$$

$$\text{s.t.} \quad \sum_{e\in E}\alpha_e\left[-k_e(\boldsymbol{s})\left(\frac{k_e(\boldsymbol{s})+\rho-1}{\rho^2}\right)+k_e(\boldsymbol{s}^*)\left(\frac{k_e(\boldsymbol{s})+\rho}{\rho^2}\right)\right]\geq 0 \quad (5.3)$$

$$\overbrace{\sum_{e\in E}\alpha_e k_e(\boldsymbol{s}^*)\left(\frac{k_e(\boldsymbol{s}^*)+\rho-1}{\rho^2}\right)}^{\text{SUM}(\boldsymbol{s}^*)}=1 \quad (5.4)$$

$$\alpha_e \geq 0, \quad \forall e\in E,$$

where (5.3) is obtained by considering the weighted sum of all the Nash constraints explicated in (5.2), over all players $i\in N$.

By taking the dual of LP, where we associate the dual variable x to the primal constraint (5.3) and the dual variable γ to the primal constraint (5.4), we get:

$$\text{DLP}: \min \ \gamma$$

$$\text{s.t.} \quad \gamma\left(\frac{k_e(\boldsymbol{s}^*)(k_e(\boldsymbol{s}^*)+\rho-1)}{\rho^2}\right)\geq k_e(\boldsymbol{s})\left(\frac{k_e(\boldsymbol{s})+\rho-1}{\rho^2}\right)$$

$$+x\left[-k_e(\boldsymbol{s})\left(\frac{k_e(\boldsymbol{s})+\rho-1}{\rho^2}\right)+k_e(\boldsymbol{s}^*)\left(\frac{k_e(\boldsymbol{s})+\rho}{\rho^2}\right)\right], \quad \forall e\in E \quad (5.5)$$

$$x\geq 0.$$

By setting $k:=k_e(\boldsymbol{s})$ and $o:=k_e(\boldsymbol{s}^*)$ in (5.5), and multiplying both sides by ρ^2, we obtain the following stronger class of dual constraints in variables x,γ:

$$xk(k+\rho-1)-xo(k+\rho)+\gamma o(o+\rho-1)\geq k(k+\rho-1), \quad \forall k,o\in\mathbb{N}_{\geq 0}. \quad (5.6)$$

To complete the proof, we are left to provide, for each $\rho\geq 1$, a suitable value $x\geq 0$ satisfying inequality (5.6) for any integers $k,o\geq 0$, where γ is set to be equal to the claimed upper bound on the ρ-uniform price of anarchy. We now proceed by case analysis.

For $\rho\leq 2$, for which we have $\gamma=\frac{5}{\rho+1}$, set $x=\frac{\rho+2}{\rho+1}$. By substituting these values in (5.6), we get the inequality $k^2-k(o(\rho+2)+\rho+1)+o(5o-\rho^2+3\rho-5)\geq 0$ which can be easily shown to be satisfied for any integers $k,o\geq 0$ when $\rho=1,2$. In fact, the discriminant of the associated equality is negative for each integer $o\geq 2$, while the cases of $o\in\{0,1\}$ can be checked by inspection.

For $\rho=3$, for which we have $\gamma=\frac{2\sqrt{7}-1}{3}$, set $x=2\sqrt{7}-4$. By substituting these values in (5.6), we get the inequality

$$(6\sqrt{7}-15)k^2-6k\left(o(\sqrt{7}-2)-2\sqrt{7}+5\right)+o\left(o(2\sqrt{7}-1)-14\sqrt{7}+34\right)\geq 0,$$

which can be easily shown to be satisfied for any integers $k, o \geq 0$. In fact, the discriminant of the associated equality is negative for each integer $o \geq 2$, while the cases of $o \in \{0,1\}$ can be checked by inspection.

For $\rho \geq 4$, for which we have $\gamma = \frac{4}{3}$, set $x = \frac{4}{3}$. By substituting these values in (5.6), we get the inequality $k^2 - k(4o - \rho + 1) + 4o(o-1) \geq 0$ whose left-hand side is increasing in ρ. Hence, we only need to prove that it gets satisfied for the case of $\rho = 4$, by which we get the inequality $k^2 - k(4o-3) + 4o(o-1) \geq 0$ which can be easily shown to be satisfied for any integers $k, o \geq 0$. Again, the discriminant of the associated equality is negative for each integer $o \geq 2$, while the cases of $o \in \{0,1\}$ can be checked by inspection. □

Now, we get matching lower bounds for each $\rho \leq 3$. We omit the case $\rho \geq 4$, as in the next subsection, we shall provide a matching lower bound holding even for the price of stability and for load balancing models (see Theorem 5.7).

Theorem 5.4. *For any $\rho \leq 3$ and $\delta > 0$, there exists an affine unweighted congestion model* $\mathsf{CM} := \mathsf{CM}(\rho, \delta)$ *such that*

$$\mathsf{PoA}_\rho(\rho\text{-}\mathsf{CG}(\mathsf{CM})) \geq \begin{cases} \frac{5}{\rho+1} & \text{if } \rho \leq 2, \\ \frac{2\sqrt{7}-1}{3} - \delta & \text{if } \rho = 3. \end{cases}$$

Proof. For $\rho = 1$, a matching lower bound has been given in [7, 12, 59]. For $\rho = 2, 3$, we simply need to define a congestion model $\widehat{\mathsf{CM}}$ whose latency functions are all derived from linear ones and such that

$$\mathsf{PoA}(\mathsf{CG}(\widehat{\mathsf{CM}})) \geq \begin{cases} \frac{5}{\rho+1} & \text{if } \rho \leq 2, \\ \frac{2\sqrt{7}-1}{3} - \delta & \text{if } \rho = 3. \end{cases}$$

Indeed, by exploiting the isomorphism of Theorem 5.1, one can easily construct, from $\widehat{\mathsf{CM}}$, a ρ-uniform congestion game CM having the same price of anarchy.

To design $\widehat{\mathsf{CM}}$, we resort to the same construction used in Theorem 4.9. Let k_1, k_2, o_1, o_2 be four non-negative integers such that:

$$(k_1, k_2, o_1, o_2) = \begin{cases} (2,1,1,1) & \text{if } \rho = 2, \\ \left(\left\lceil \frac{\sqrt{7}+4}{3} t \right\rceil, 1, t, 1\right) & \text{if } \rho = 3, \end{cases}$$

where t is an arbitrary non-negative integer.

Let $\widehat{\mathsf{CM}}(k_1, k_2, o_1, o_2)$ be a general congestion model defined as follows:

- we have a set N of $n := \max\{k_1 + o_1, k_2 + o_2\}$ players, and a set of resources $E := E_1 \cup E_2$, where $E_j := \{e_{j,1}, \ldots, e_{j,n}\}$ for any $j \in [2]$;
- each player $i \in N$ has two strategies s_i, s_i^* only, namely,

$$s_i := \{e_{1,i}, \ldots, e_{1,i+k_1-1}\} \cup \{e_{2,i}, \ldots, e_{2,i+k_2-1}\} \text{ and}$$
$$s_i^* := \{e_{1,i+k_1}, \ldots, e_{1,i+k_1+o_1-1}\} \cup \{e_{2,i+k_2}, \ldots, e_{2,i+k_2+o_2-1}\},$$

where the resources' indexes are cyclical (i.e., $e_{j,i+n} := e_{j,i}$);

- for any $j \in [2]$, the latency function of each resource $e \in E_j$ is defined as $\ell_j(x) = \alpha_j \cdot g(x)$ for any $x \geq 0$, where $g(x) := \frac{x+\rho-1}{\rho^2}$ for any $x \geq 0$, $\alpha_1 := o_2 \cdot g(k_2+1) - k_2 \cdot g(k_2)$, and $\alpha_2 := k_1 \cdot g(k_1) - o_1 \cdot g(k_1+1)$.

Observe that, by Lemma 5.1, $\ell_j(x)$ is derived according to ρ from $\alpha_j x$ for any $j \in [2]$, thus the latency functions of $\widehat{\mathsf{CM}}$ are all derived from linear ones.

Let $\boldsymbol{s}$ and $\boldsymbol{s}^*$ be the strategy profiles in which each player $i \in N$ plays strategy s_i and s_i^*, respectively. We observe that $k_e(\boldsymbol{s}) = k_j$, $k_e(\boldsymbol{s}^*) = o_j$ and $k_e(\boldsymbol{s}_{-i}, s_i^*) = k_j + 1$, for any resource e of group E_j (with $j \in [2]$) and for any player $i \in N$; furthermore, the cost of any player $i \in N$ in $\boldsymbol{s}$ (resp. $\boldsymbol{s}^*$, resp. $(\boldsymbol{s}_{-i}, s_i^*)$) is $\alpha_1 k_1 g(k_1) + \alpha_2 k_2 g(k_2)$ (resp. $\alpha_1 o_1 g(o_1) + \alpha_2 o_2 g(o_2)$, resp. $\alpha_1 o_1 g(k_1+1) + \alpha_2 o_2 g(k_2+1)$). As

$$cost_i(\boldsymbol{s}) = \alpha_1 k_1 g(k_1) + \alpha_2 k_2 g(k_2) = \alpha_1 o_1 g(k_1+1) + \alpha_2 o_2 g(k_2+1) = cost_i(\boldsymbol{s}_{-i}, s_i^*)$$

for any player $i \in N$, we have that $\boldsymbol{s}$ is a pure Nash equilibrium of $\mathsf{CG}(\widehat{\mathsf{CM}})$.

If $\rho = 2$, we have that

$$\frac{\mathsf{SUM}(\boldsymbol{s})}{\mathsf{SUM}(\boldsymbol{s}^*)} = \frac{(\alpha_1 k_1 g(k_1) + \alpha_2 k_2 g(k_2))n}{(\alpha_1 o_1 g(o_1) + \alpha_2 o_2 g(o_2))n} = \frac{5}{3},$$

thus

$$\mathsf{PoA}(\mathsf{CG}(\widehat{\mathsf{CM}}(k_1,k_2,o_1,o_2))) = \frac{5}{3}.$$

Analogously, if $\rho = 3$, we have that the ratio $\frac{\mathsf{SUM}(\boldsymbol{s})}{\mathsf{SUM}(\boldsymbol{s}^*)}$ approaches

$$\frac{(\alpha_1 k_1 g(k_1) + \alpha_2 k_2 g(k_2))n}{(\alpha_1 o_1 g(o_1) + \alpha_2 o_2 g(o_2))n} = \frac{2\sqrt{7}-1}{3}$$

for t going to infinite (we recall that the values of k_1 and o_1 depends on t, if $\rho = 3$). Thus, by choosing a sufficiently large t, we get

$$\mathsf{PoA}(\mathsf{CG}(\widehat{\mathsf{CM}}(k_1,k_2,o_1,o_2))) \geq \frac{2\sqrt{7}-1}{3} - \delta,$$

and this concludes the proof. □

Remark 5.1. The choice of parameters k_1, k_2, o_1, o_2 in the proof of Theorem 5.4 has been "suggested" by the primal-dual approach. Indeed, consider the class of dual constraints (5.6) used to show the upper bounds of Theorem 5.3, and let x, γ be the (optimal) values of the dual variables considered in the related proof. For $\rho = 2$ (resp. $\rho = 3$) the above dual constraint is tight (resp. asymptotically tight for $t \to \infty$) if $(k,o) = (k_1, o_1)$, and it is also tight if $(k,o) = (k_2, o_2)$. Thus, by adopting a similar approach as in Example 3.1, we are able to construct a simplified version of the primal program LP as that considered in Theorem 5.3, but with two resources only, such that $k_1(\boldsymbol{s}) = k_1$, $k_1(\boldsymbol{s}^*) = o_1$, $k_2(\boldsymbol{s}) = k_2$, $k_2(\boldsymbol{s}^*) = o_2$, and such that the primal constraint (5.3) is tight. Such a linear program has "suggested" the exact structure of the lower bounding instance $\widehat{\mathsf{CM}}(k_1,k_2,o_1,o_2)$.

The Price of Anarchy for General Latency Functions

In this subsection, we give the main intuitions on how to determine the price of anarchy for general unweighted congestion models, thus extending the results obtained for affine latency functions. To this aim, we resort to the machinery presented in Chapter 4.

We observe that Theorem 5.1 establishes an isomorphism (via map Λ) between games ρ-$\mathsf{CG}(\mathsf{CM})$ and $\mathsf{CG}(\Lambda(\mathsf{CM},\rho))$, and allows us to treat any ρ-uniform unweighted congestion game with latency functions in a given class $\mathcal{C}$, as a classical unweighted games whose latency functions are derived according to ρ from latencies in $\mathcal{C}$. Hence, we can exploit the results of Theorem 4.9 to get tight upper and lower bounds on the price of anarchy for ρ-uniform unweighted congestion games with general latency functions. In particular, we get that the tight bound on the price of anarchy of ρ-uniform unweighted games with latency functions in $\mathcal{C}$ is equal to $\gamma_{\mathsf{U}}(\mathsf{PoA}_\varepsilon,\mathcal{C}_\rho)$, where γ_{U} is defined as in Section 4.5 and $\mathcal{C}_\rho := \{\lambda_{\ell,\rho} : \ell \in \mathcal{C}\}$ is the set of latencies derived from $\mathcal{C}$ according to ρ. Furthermore, the map Λ can be easily (and slightly) modified to preserve the load balancing structure of the game. Then, by applying Theorem 4.7, one can provide tight lower bounds holding even for load balancing games, if the considered latency functions are closed under ordinate-scaling.

We point out that the results obtained for the case of affine latencies in the previous section can be reinterpreted in terms of this general framework. Indeed, the upper bound provided in Theorem 5.3 coincides with $\gamma_{\mathsf{U}}(\mathsf{PoA}_\varepsilon,\mathcal{P}(1)_\rho)$, so that the lower bounding instance is automatically guaranteed by our findings (which hold for more general latency functions). We also observe that the lower bounding instance provided in Theorem 5.4 coincides with the lower bounding instance of Theorem 4.9, in which the opportune choice of the parameters k_1,k_2,o_1,o_2,f_1,f_2 satisfying Lemma 4.4 is that considered in the proof of Theorem 5.4 (with f_1 and f_2 being the latency functions derived from the $f(x)=x$, according to ρ).

Price of Stability for Affine Functions

In this subsection, we exhibit exact bounds on the price of stability of ρ-uniform congestion games induced by affine unweighted congestion models.

Theorem 5.5. *Fix an affine unweighted congestion model* CM. *For any* $\rho \geq 1$, *we have*

$$\mathsf{PoS}_\rho(\rho\text{-}\mathsf{CG}(\mathsf{CM})) \leq \begin{cases} 1+\frac{1}{\sqrt{2\rho+1}} & \text{if } \rho \leq 3, \\ \frac{4}{3} & \text{if } \rho \geq 4. \end{cases}$$

Proof. Given an integer $\rho \geq 1$, let CM be an arbitrary affine unweighted congestion model. Again, we can assume without loss of generality that the latency functions are of type $\ell_e(x) = \alpha_e x$ (i.e., linear). As $\mathsf{PoA}_\rho(\rho\text{-}\mathsf{CG}(\mathsf{CM})) = \mathsf{PoA}(\mathsf{CG}(\Lambda(\mathsf{CM},\rho)))$ (by Theorem 5.1), it is sufficient showing that the desired upper bound holds for $\mathsf{PoA}(\mathsf{CG}(\Lambda(\mathsf{CM},\rho)))$. We recall that, by Rosenthal's Theorem [154], the function Φ defined as

$$\Phi(\boldsymbol{s}) := \sum_{e\in E}\sum_{j=1}^{k_e(\boldsymbol{s})} \lambda_{\ell_e,\rho}(k_e(\boldsymbol{s})) = \sum_{e\in E} \alpha_e \frac{k_e(\boldsymbol{s})\,(k_e(\boldsymbol{s})+2\rho-1)}{2\rho^2}$$

is a potential function of $\mathsf{CG}(\Lambda(\mathsf{CM},\rho))$, where the second equality follows from Lemma 5.1; thus, any global minimum of Φ is a pure Nash equilibrium. Let $\boldsymbol{s}$ and $\boldsymbol{s}^*$ be a pure Nash equilibrium minimizing the potential function Φ and a social optimum for $\mathsf{CG}(\Lambda(\mathsf{CM},\rho))$, respectively, so that the following inequality holds: $\Phi(\boldsymbol{s}) \leq \Phi(\boldsymbol{s}^*)$. By definition of Φ, this inequality can be equivalently written as

$$\sum_{e\in E} \alpha_e \left[-k_e(\boldsymbol{s})\left(\frac{k_e(\boldsymbol{s})+2\rho-1}{2\rho^2}\right) + k_e(\boldsymbol{s}^*)\left(\frac{k_e(\boldsymbol{s}^*)+2\rho-1}{2\rho^2}\right)\right] \geq 0.$$

By adding it to the formulation used within the proof of Theorem 5.3 for upper bounding the price of anarchy, we get the following linear program:

$$\mathsf{LP}:\ \max \sum_{e\in E} \alpha_e \overbrace{\left(\frac{k_e(\boldsymbol{s})(k_e(\boldsymbol{s})+\rho-1)}{\rho^2}\right)}^{\mathsf{SUM}(\boldsymbol{s})}$$

$$\text{s.t.} \sum_{e\in E} \alpha_e \overbrace{\left[-k_e(\boldsymbol{s})\left(\frac{k_e(\boldsymbol{s})+\rho-1}{\rho^2}\right) + k_e(\boldsymbol{s}^*)\left(\frac{k_e(\boldsymbol{s})+\rho}{\rho^2}\right)\right]}^{=:A(\boldsymbol{s},\boldsymbol{s}^*)} \geq 0 \qquad (5.7)$$

$$\sum_{e\in E} \alpha_e \overbrace{\left[-k_e(\boldsymbol{s})\left(\frac{k_e(\boldsymbol{s})+2\rho-1}{2\rho^2}\right) + k_e(\boldsymbol{s}^*)\left(\frac{k_e(\boldsymbol{s}^*)+2\rho-1}{2\rho^2}\right)\right]}^{=:B(\boldsymbol{s},\boldsymbol{s}^*)} \geq 0 \qquad (5.8)$$

$$\sum_{e\in E} \alpha_e \overbrace{\left(\frac{k_e(\boldsymbol{s}^*)(k_e(\boldsymbol{s}^*)+\rho-1)}{\rho^2}\right)}^{\mathsf{SUM}(\boldsymbol{s}^*)} = 1 \qquad (5.9)$$

$$\alpha_e \geq 0, \quad \forall e \in E.$$

By taking the dual of LP, where we associate the dual variable x to the primal constraint (5.7), the dual variable y to the primal constraint (5.8), and the dual variable

γ to the primal constraint (5.9), we get

$$\begin{aligned} \mathsf{DLP}: \quad & \min \quad \gamma \\ & \text{s.t.} \ \gamma \cdot \mathsf{SUM}(\boldsymbol{s}^*) \geq \mathsf{SUM}(\boldsymbol{s}) + x \cdot A(\boldsymbol{s},\boldsymbol{s}^*) + y \cdot B(\boldsymbol{s},\boldsymbol{s}^*), \ \forall e \in E \quad (5.10) \\ & \quad\ x, y \geq 0. \end{aligned}$$

By using $x_i = x$ for each $i \in N$, $k := k_e(\boldsymbol{s})$ and $o := k_e(\boldsymbol{s}^*)$ in (5.10), and multiplying both sides by ρ^2, we obtain the following stronger class of dual constraints:

$$\begin{aligned} \gamma o(o+\rho-1) \geq {} & k(k+\rho-1) + x[-k(k+\rho-1) + o(k+\rho)] \\ & + y\left[-\frac{k(k+2\rho-1)}{2} + \frac{o(o+2\rho-1)}{2}\right], \quad \forall k, o \in \mathbb{N}_{\geq 0} \end{aligned} \quad (5.11)$$

For $\rho \geq 4$, the claim follows by Theorem 5.3. For each $\rho \leq 3$, as done in Theorem 5.3 for the price of anarchy, there are suitable $x, y \geq 0$ satisfying inequality (5.11) for any integers $k, o \geq 0$, where γ is set to be equal to the claimed upper bound on the ρ-uniform price of stability. In particular, such values are $\gamma = 1 + \frac{1}{\sqrt{2\rho+1}}$, $x = \frac{\rho}{\sqrt{2\rho+1}}$ and $y = 1 - \frac{\rho-1}{\sqrt{2\rho+1}}$. □

In the proof of Theorem 5.3, as the dual variables are two (i.e. x, γ), their optimal values are determined by finding two pairs (k_1, o_1) and (k_2, o_2) for which the dual constraints are tight. Instead, in the proof of Theorem 5.5 (for $\rho \leq 3$), the dual variables are three (i.e., x, y, γ), and their optimal values can be determined by finding three distinct pairs $(k_1, o_1), (k_2, o_2)$ and (k_3, o_3). These three pairs again suggest how to design a congestion game that provides a lower bound, and the considered high-level structure is similar to those used in [60]. Thus, we get the following theorem.

Theorem 5.6. *For each $\rho \leq 3$ and $\delta > 0$, there exists an affine unweighted congestion model* $\mathsf{CM} := \mathsf{CM}(\rho, \delta)$ *such that* $\mathsf{PoS}_\rho(\rho\text{-}\mathsf{CG}(\mathsf{CM})) \geq 1 + \frac{1}{\sqrt{2\rho+1}} - \delta$.

For $\rho \geq 4$, both the price of anarchy and the price of stability are upper bounded by $4/3$, that is by the price of anarchy of affine non-atomic games. Thus, to show tight bounds, one can simply transform the tight lower bound instance for classical affine non-atomic congestion games, that is a simple symmetric load balancing game, into a ρ-uniform congestion game having the same price of anarchy, and the same combinatorial structure of the players' strategic space. Then, the following theorem easily follows.

Theorem 5.7. *For each $\rho \geq 1$ and $\delta > 0$, there exists an affine unweighted symmetric load balancing model* $\mathsf{CM} := \mathsf{CM}(\rho, \delta)$ *such that* $\mathsf{PoS}_\rho(\rho\text{-}\mathsf{CG}(\mathsf{CM})) \geq \frac{4}{3} - \delta$.

5.5 Concluding Remarks and Related Work

The results presented in this chapter are based on the work of Bilò et al. [35]. There, together with the results for unweighted games, it is also analyzed the existence of equilibria and the efficiency of the more general ρ-uniform weighted congestion games. In particular, Bilò et al. [35] prove that ρ-uniform mixed equilibria do exist in affine weighted congestion games, for each $\rho \geq 1$. This generalizes to every value of ρ the results in Fotakis et al. [90], Harks [100] and Panagopoulou and Spirakis [142] which were proved for the classical setting in which agents adopt pure strategies, i.e., for $\rho = 1$. Then, for each $\rho \geq 1$, by exploiting the primal-dual method, they derive tight bounds on the price of anarchy of ρ-uniform mixed equilibria in affine weighted congestion games (see the values reported in Table 5.1, where the lower bounds for weighted games hold even for parallel-link networks). It is worth noticing that such results nicely extend the ones previously obtained in the literature for pure Nash equilibria, i.e., the case of $\rho = 1$.

Analogously to ρ-uniform congestion games, there are other research works in which affine congestion games have been generalised according to a certain probabilistic framework (see, for instance, Pilouras et al. [151], Bilò and Vinci [28], and Cominetti et al. [75]). In the work of Piliouras et al. [151], the price of anarchy of affine unweighted congestion games in which the users of a resource are scheduled in a random order and the cost of each agent is defined as the sum of her completion time on each selected resource is evaluated. These games happen to coincide with our model with $\rho = 2$. Thus, our findings provide a tight characterization of the price of stability for their model which was not investigated under this best-case metric. Bilò and Vinci [28] considered a generalization of the model defined by Piliouras et al. [151], in which users generally belong to different priority classes, and the users belonging to the same class and selecting the same resource are again scheduled in a random order.

In the work of Cominetti et al. [75], which appeared after the publication of [35], the price of anarchy of affine unweighted congestion games in which agents may participate according to a certain probability is characterized. In this case, the model presented in this chapter happens to coincide with their games when this probability is of the form $1/\rho$, for any positive integer ρ. A general cost model for congestion games that can capture in a unified manner all these phenomena has been put forward by Kleer and Schäfer [109], and bounds on the price of anarchy and the price of stability have been given for the case of affine unweighted games. Another general congestion model in which the agents assigned to each resource are scheduled randomly has been introduced and studied by Penn et al. [146, 147].

Ashlagi et al. [9] investigate symmetric affine congestion games under the assumption that the number of participating agents may be unknown. They consider the notion of safety-level equilibrium defined by Aghassi and Bertsimas [5] and show that agents may benefit from the common ignorance about the number of participants.

Chapter 6
Non-atomic Congestion Games with θ-Similar Strategies

What route should I choose to go to work? This is a fundamental question that millions of people ask themselves daily, especially in metropolitan areas where the set of possible alternatives may be considerably diversified. In an ideal situation in which travel time is not affected by traffic, everyone would select the fastest path. However, when some roads become heavily congested, the ideal travel time may tremendously increase and some people may prefer taking longer detours to avoid traffic delay. As every worker can be seen as a selfish and rational agent who is only interested in minimizing its travel time, this type of problems gets suitably modeled as non-cooperative strategic games, usually called *routing games*, and coincide with the model of non-atomic congestion games defined in Chapter 2.

In this chapter, we focus on the well-studied case of affine functions, and investigate to what extent the price of anarchy is affected by the similarities among the players' strategies. Our notion of similarity is modeled by assuming that, given a parameter $\theta \geq 1$, the costs of any two strategies available to a same player, when evaluated in absence of congestion, are within a factor θ one from the other. In particular, we show that the price of anarchy of non-atomic games with affine functions and θ-similar strategies is exactly $\frac{\left(\sqrt{\theta(\theta-1)}+\theta-2\right)^2}{(3\theta-4)\left(2\sqrt{\theta(\theta-1)}-\theta\right)}$ if $\theta \neq \frac{4}{3}$ and $\frac{9}{8}$ for $\theta = \frac{4}{3}$. Hence, as $\frac{\left(\sqrt{\theta(\theta-1)}+\theta-2\right)^2}{(3\theta-4)\left(2\sqrt{\theta(\theta-1)}-\theta\right)}$ is increasing in θ and approaches $\frac{4}{3}$ when θ goes to infinity, improvements on the $\frac{4}{3}$-bound shown by Roughgarden and Tardos [159] can be obtained for any finite value of θ. Furthermore, we show that the price of anarchy of simple Pigou's networks (i.e., symmetric load balancing games with two resources only) with θ-similar strategies drops to $\frac{4\theta}{3\theta+1}$ and coincides with that of general games only if $\theta \in \{1, \infty\}$. See Figure 6.1 for some numerical comparisons.

V. Bilò, C. Vinci, *Coping with Selfishness in Congestion Games*, Monographs in Theoretical Computer Science. An EATCS Series, https://doi.org/10.1007/978-3-031-30261-9_6

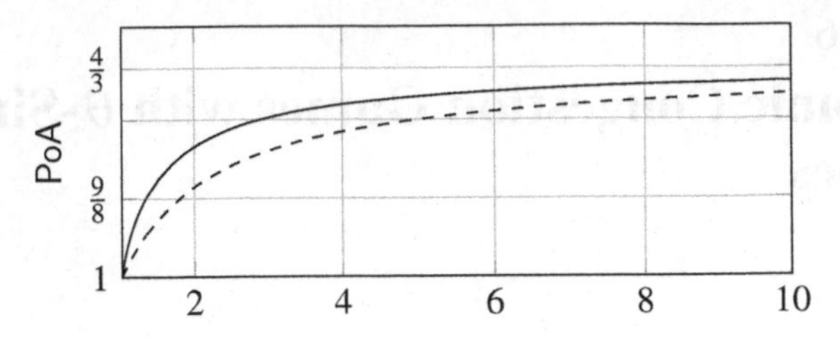

Fig. 6.1 The price of anarchy of non-atomic affine congestion games as a function of θ, in both the general case (solid line) and Pigou's networks (dashed line).

6.1 Model and Definitions

For general non-atomic games, we recall the model and the related concepts (e.g., pure Nash equilibria, price of anarchy, etc.) from Chapter 2 (in particular, from Section 2.5).

Definition 6.1 (Congestion Game with θ-similar strategies). Given $\theta \geq 1$, a *non-atomic congestion game with θ-similar strategies* $\mathsf{NCG}(\theta)$ is a congestion game in which, for each $i \in N$ and strategies $S, S' \in \Sigma_i$, it holds that

$$\sum_{e \in S} \ell_e(0) \leq \theta \sum_{e \in S'} \ell_e(0),$$

i.e., all the strategies available to players of type i, when evaluated in absence of congestion, are within a factor θ one from the other.

Observe that congestion games with θ-similar strategies are congestion games obeying some special properties. Thus, all positive results holding for congestion games carry over to such variants for any value of θ. Moreover, for $\theta = \infty$, any congestion game is a congestion game with θ-similar strategies.

6.2 PoA for General Games

In this section, by applying the primal-dual method, we derive bounds on the price of anarchy of non-atomic congestion games with θ-similar strategies, for any $\theta \geq 1$, without any restriction on the topology of the game, i.e., the players' strategic space is general. We start by showing a general upper bound.

Theorem 6.1. *Let* $\mathsf{NCG}(\theta)$ *be an arbitrary affine non-atomic congestion game with θ-similar strategies. Then:*

$$\mathsf{PoA}(\mathsf{NCG}(\theta)) \leq \begin{cases} \frac{\left(\sqrt{(\theta-1)\theta}+\theta-2\right)^2}{(3\theta-4)\left(2\sqrt{(\theta-1)\theta}-\theta\right)} & \text{if } \theta \neq \frac{4}{3}, \\ \frac{9}{8} & \text{if } \theta = \frac{4}{3}. \end{cases}$$

Proof. Fix an arbitrary affine non-atomic congestion game with θ-similar strategies

$$\mathsf{NCG}(\theta) := (N, E, (\ell_e)_{e\in E}, (r_i)_{i\in N}, (\Sigma_i)_{i\in N}).$$

Assume that $1 \le \theta < \frac{4}{3}$ or $\frac{4}{3} < \theta$. Let $\boldsymbol{\sigma} := (\sigma_{i,S})_{i\in N, S\in\Sigma_i}$ and $\boldsymbol{\sigma}^* := (\sigma^*_{i,S})_{i\in N, S\in\Sigma_i}$ be a pure Nash equilibrium and a social optimum of $\mathsf{NCG}(\theta)$, respectively. As the latency functions are affine, we can set $\ell_e(x) := \alpha_e x + \beta_e$ for any resource $e \in E$, for some $\alpha_e, \beta_e \ge 0$. As usual, we denote $k_e(\boldsymbol{\sigma})$ and $k_e(\boldsymbol{\sigma}^*)$ with k_e and o_e, respectively. Observe that we can assume without loss of generality that all players of the same type play the same strategy S_i in $\boldsymbol{\sigma}$, and play the same strategy S_i^* in $\boldsymbol{\sigma}^*$, so that $\sigma_{i,S_i} = \sigma^*_{i,S_i^*}$. Indeed, if it is not the case, it is sufficient splitting the players of each type in several sub-types in such a way that the above property holds. By applying the primal-dual method, we get the following linear program in variables α_es and β_es, whose optimal value is an upper bound on $\mathsf{PoA}(\mathsf{NCG}(\theta))$:

$$\mathsf{LP}: \quad \max \sum_{e\in E} (\alpha_e k_e^2 + \beta k_e) \tag{6.1}$$

$$\text{s.t.} \quad \sum_{e\in E} (\alpha_e k_e^2 + \beta k_e) \le \sum_{e\in E} (\alpha_e o_e k_e + \beta_e o_e) \tag{6.2}$$

$$\sum_{e\in E} \beta_e o_e \le \theta \sum_{e\in E} \beta_e k_e \tag{6.3}$$

$$\sum_{e\in E} (\alpha_e o_e^2 + \beta_e o_e) = 1 \tag{6.4}$$

$$\alpha_e, \beta_e \ge 0, \quad \forall e \in E.$$

In particular:

- the objective function (6.1) is the total latency $\mathsf{SUM}(\boldsymbol{\sigma})$;
- (6.2) has been obtained by

$$\begin{aligned} 0 &\ge \sum_{i\in N} \sigma_{i,S_i} \left(c_{S_i}(\boldsymbol{\sigma}) - c_{S_i^*}(\boldsymbol{\sigma}) \right) \\ &= \sum_{i\in N} \sigma_{i,S_i} \cdot c_{S_i}(\boldsymbol{\sigma}) - \sum_{i\in N} \sigma^*_{i,S_i^*} \cdot c_{S_i^*}(\boldsymbol{\sigma}) \\ &= \sum_{i\in N} \sigma_{i,S_i} \sum_{e\in S_i} (\alpha_e k_e + \beta_e) - \sum_{i\in N} \sigma^*_{i,S_i^*} \sum_{e\in S_i^*} (\alpha_e k_e + \beta_e) \\ &= \sum_{e\in E} (\alpha_e k_e^2 + \beta_e k_e) - \sum_{e\in E} (\alpha_e o_e k_e + \beta_e o_e), \end{aligned} \tag{6.5}$$

  where (6.5) holds since $\boldsymbol{\sigma}$ is a pure Nash equilibrium, and then $c_{S_i}(\boldsymbol{\sigma}) - c_{S_i^*}(\boldsymbol{\sigma}) \le 0$ holds for any $i \in N$ such that $\sigma_{i,S_i} > 0$;
- (6.3) has been obtained by

$$0 \ge \sum_{i\in N} \sigma_{i,S_i} \left(\sum_{e\in S_i^*} \ell_e(0) - \theta \sum_{e\in S_i} \ell_e(0) \right) \tag{6.6}$$

$$= \sum_{i \in N} \sigma^*_{i,S^*_i} \sum_{e \in S^*_i} \ell_e(0) - \theta \sum_{i \in N} \sigma_{i,S_i} \sum_{e \in S_i} \ell_e(0)$$
$$= \sum_{i \in N} \sigma^*_{i,S^*_i} \sum_{e \in S^*_i} \beta_e - \theta \sum_{i \in N} \sigma_{i,S_i} \sum_{e \in S_i} \beta_e$$
$$= \sum_{e \in E} \beta_e o_e - \theta \sum_{e \in E} \beta_e k_e,$$

where (6.6) holds since the game is with θ-similar strategies;

- the linear coefficients α_es and β_es are the variables of the linear program, and the other quantities are fixed parameters;
- (6.4) normalizes the optimal total latency $\mathsf{SUM}(\boldsymbol{\sigma}^*)$, so that, analogously to Example 3.1, the maximum value of the objective function is an upper bound on the price of anarchy.

By taking the dual of LP, in which we associate the dual variable x to the primal constraint (6.2), the dual variable y to the primal constraint (6.3), and the dual variable γ to the primal constraint (6.4), we get:

$$\begin{aligned}
&\mathsf{DLP}: \\
&\quad \min \quad \gamma \\
&\quad \text{s.t.} \quad \gamma o_e^2 \geq k_e^2 + x\left(-k_e^2 + o_e k_e\right) \quad \forall e \in E && (6.7)\\
&\quad \quad \quad \gamma o_e \geq k_e + x(-k_e + o_e) + y(-o_e + \theta k_e) \geq 0, \quad \forall e \in E && (6.8)\\
&\quad \quad \quad x, y \geq 0.
\end{aligned}$$

We show that, by setting

$$\gamma := \gamma(\theta) := \frac{\left(\sqrt{(\theta-1)\theta} + \theta - 2\right)^2}{(3\theta - 4)\left(2\sqrt{(\theta-1)\theta} - \theta\right)},$$
$$x := x(\theta) := \frac{2\sqrt{\theta^2 - \theta} + 2\theta - 4}{3\theta - 4} \geq 1,$$
$$y := y(\theta) := \frac{x-1}{\theta} \geq 0,$$

all the constraints in (6.7) and (6.8) are verified for any $k_e, o_e \in \mathbb{R}_{\geq 0}$. First of all, since $x \geq 1$ and $y \geq 0$, we trivially get that the dual constraints are satisfied by $o_e = 0$. Thus, we just consider these constraints with $o_e > 0$. By rewriting DLP in terms of variable $t := k_e / o_e \geq 0$, and by eliminating variable y, we get the following stronger class of dual constraints:

$$\gamma \geq t^2 + x\left(-t^2 + t\right), \; \forall t \geq 0 \tag{6.9}$$
$$\gamma \geq (1 - x + y\theta)t + x - y \geq 0, \; \forall t \geq 0, \tag{6.10}$$

and one can easily prove that these inequalities are satisfied for any $t > 0$, thus showing the claim.

Finally, if $\theta = \frac{4}{3}$, it suffices considering the limit values, for θ tending to $\frac{4}{3}$, of all the quantities defined above, and the claim holds as well. □

Remark 6.1. The values assigned to x, y and γ (in the proof of Theorem 6.1) have been chosen in such a way that all the constraints of type (6.7) and (6.8) are satisfied, and such that γ is minimized. Furthermore, by using this assignment, we have that constraint (6.8) is tight if $(k_e, o_e) = (1,0)$ or $(k_e, o_e) = (0,1)$, and constraint (6.7) is tight if $(k_e, o_e) = (\xi, 1)$, where ξ is defined as

$$\xi := \xi(\theta) := \begin{cases} \frac{\sqrt{\theta^2-\theta}+\theta-2}{2\sqrt{\theta^2-\theta}-\theta}, \text{ if } \theta \neq \frac{4}{3}, \\ \frac{3}{2}, \text{ if } \theta = \frac{4}{3}. \end{cases}$$

By exploiting these facts, we also obtain that the considered assignment for the variables x, y and γ is the optimal solution of a dual program as in the proof of Theorem 6.1, but with a unique constraint of type (6.7) parametrized by $(k_e, o_e) = (\xi, 1)$, and two constraints of type (6.8) parametrized respectively by $(k_e, o_e) = (1,0)$ and $(k_e, o_e) = (0,1)$:

$$\begin{aligned} \overline{\mathsf{DLP}}: \quad \min \quad & \gamma \\ \text{s.t.} \quad & \gamma \geq \xi^2 + x(-\xi^2 + \xi) \\ & 0 \geq 1 - x + y\theta \\ & \gamma \geq x - y \\ & x, y \geq 0. \end{aligned}$$

Thus, by applying the Strong Duality Theorem to $\overline{\mathsf{DLP}}$, we get that the value assigned to γ (i.e., the upper bound shown in Theorem 6.1) is the optimal solution of a simplified version of the primal program defined in Theorem 6.1, but with three variables α, β_1, β_2 only (associated respectively to the three constraints of $\overline{\mathsf{DLP}}$), and such that the primal constraints are tight (this fact holds by the complementary slackness conditions, as both x and y are strictly positive). Thus, we get the following linear program in variables α, β_1, β_2:

$$\begin{aligned} \overline{\mathsf{LP}}: \quad \max \quad & \alpha\xi^2 + \beta_1 \\ \text{s.t.} \quad & \alpha\xi^2 + \beta_1 = \alpha\xi + \beta_2 \\ & \beta_2 = \theta\beta_1 \\ & \alpha + \beta_2 = 1 \\ & \alpha, \beta_1, \beta_2 \geq 0. \end{aligned}$$

Furthermore, the optimal assignments for α, β_1, β_2 are univocally determined by the three tight constraints of $\overline{\mathsf{LP}}$, and such values have suggested the structure of a tight lower bound.

Theorem 6.2. *For any $\theta \geq 1$ and for any $\delta > 0$, there exists an affine non-atomic congestion game with θ-similar strategies* $\mathsf{NCG}(\theta)$ *such that:*

$$\mathsf{PoA}(\mathsf{NCG}(\theta)) \geq \begin{cases} \frac{\left(\sqrt{(\theta-1)\theta}+\theta-2\right)^2}{(3\theta-4)\left(2\sqrt{(\theta-1)\theta}-\theta\right)} - \delta & \text{if } \theta \neq \frac{4}{3}, \\ \frac{9}{8} - \delta & \text{if } \theta = \frac{4}{3}. \end{cases}$$

Proof. Fix $\theta \geq 1$ and let $\xi := \xi(\theta)$ be the quantity defined in Remark 6.1. For simplicity, we first assume that both θ and ξ are rational numbers. Let $(\alpha, \beta_1, \beta_2)$ be the optimal solution of the linear program $\overline{\mathsf{LP}}$ considered in Remark 6.1.

We shall construct a non-atomic congestion game with θ-similar strategies NCG that is parametrized by α, β_1, β_2, and whose price of anarchy coincides with the optimal value of $\overline{\mathsf{LP}}$. Thus, by the considerations done in Remark 6.1, it will automatically match the upper bound of Theorem 6.1[1].

Let NCG be a non-atomic congestion game defined as follows:

- let h be a non-negative integer such that ξh is integer, too;
- we have a set N of $n := (\xi + 1)h$ types of players, each one having an amount of flow $r_i = \frac{1}{h^2}$, and a set of $2\xi h + 2$ resources $E := E' \cup \{e_1, e_2\}$, with $E' = \{e'_1, \ldots, e'_n\}$;
- each player of type i has two strategies σ_i, σ_i^* only, defined as

$$s_i := \{e'_i, \ldots, e'_{i+\xi h-1}\} \cup \{e_1\}, \quad s_i^* := \{e'_{i+\xi h}, \ldots, e'_{i+\xi h+h-1}\} \cup \{e_2\},$$

  where the resources' indexes are supposed to be cyclical (i.e., $e_{j,i+n} := e_{j,i}$);
- the latency function of resources $e' \in E'$ is defined as $\ell'(x) = \alpha \cdot x$, the latency of each resource e_j is defined as $\ell_j(x) = \beta_j$, for $j \in [2]$.

We observe that NCG is a congestion games with θ-similar strategies. Indeed, by exploiting the second constraint of $\overline{\mathsf{LP}}$, we have that

$$\sum_{e \in s_i^*} \ell_e(0) = \beta_2 = \theta \beta_1 = \theta \sum_{e \in s_i} \ell_e(0)$$

for any player of type i.

Let $\boldsymbol{\sigma}$ and $\boldsymbol{\sigma}^*$ be the strategy profiles in which each player of type i plays strategy s_i and s_i^* respectively. We have that $\boldsymbol{\sigma}$ is a pure Nash equilibrium. Indeed, by exploiting the first constraint of $\overline{\mathsf{LP}}$, we get

$$cost_i(\boldsymbol{\sigma}) = \sum_{e \in s_i} \ell_e(k_e(\boldsymbol{\sigma})) = \xi h \alpha \left(\frac{1}{h}^2 \xi h\right) + \beta_1 = \alpha \xi^2 + \beta_1$$

$$= \alpha \xi + \beta_2 = h \alpha \left(\frac{1}{h}^2 \xi h\right) + \beta_2 = \sum_{e \in s_i} \ell_e(k_e(\boldsymbol{\sigma})) = cost_i(\boldsymbol{\sigma}_{-i}, s_i^*).$$

By the above equalities, we get

[1] Interestingly, in the remainder of the proof, we will not use the numerical representations of α, β_1, β_2, but their qualitative properties only (e.g., the fact that they satisfy all the constraints in $\overline{\mathsf{LP}}$). This type of reasoning is often provided by the primal-dual approach, when deriving tight lower bounds.

$$cost_i(\boldsymbol{\sigma}) = \alpha\xi^2 + \beta_1.$$

Furthermore, by exploiting the third constraint of $\overline{\mathsf{LP}}$ we get

$$cost_i(\boldsymbol{\sigma}^*) = \sum_{e \in s_i} \ell_e(k_e(\boldsymbol{\sigma}^*)) = h\alpha\left(\frac{1}{h}^2 h\right) + \beta_2 = \alpha + \beta_2 = 1.$$

Thus, we have that

$$\mathsf{PoA}(\mathsf{NCG}) \geq \frac{\frac{1}{h^2}((\xi+1)h)cost_i(\boldsymbol{\sigma})}{\frac{1}{h^2}((\xi+1)h)cost_i(\boldsymbol{\sigma}^*)} = \alpha\xi^2 + \beta_1.$$

As the last member of the inequalities coincides with the optimal value of $\overline{\mathsf{LP}}$, by Remark 6.1, we have that $\mathsf{PoA}(\mathsf{NCG})$ matches the upper bound of Theorem 6.1.

Finally, if the quantities θ and $\xi := \xi(\theta)$ used in the proof are not rational, it is sufficient considering, in place of θ, another value $\theta' < \theta$ sufficiently close to θ, such that both θ' and $\xi := \xi(\theta')$ are rational. By constructing the lower bound instance according to these new values, the θ-similarity property is preserved, and the resulting price of anarchy is equal to the upper bound of Theorem 6.1 up to a negligible additive factor (as θ' is sufficiently close to θ). □

6.3 The Case of Pigou's Networks

In this section, we provide the price of anarchy of θ-similar non-atomic congestion games with affine functions, when the topology is a *Pigou's network* [150], that is equal to a symmetric load balancing game with two resources only, and then, it is the simplest possible topology. We have the following theorem:

Theorem 6.3. *Let* $\mathsf{LB}(\theta)$ *be an arbitrary symmetric load balancing game with two resources, affine latencies, and* θ*-similar strategies. Then* $\mathsf{PoA}(\mathsf{LB}) \leq \frac{4\theta}{3\theta+1}$.

Proof. Fix an arbitrary symmetric load balancing game $\mathsf{LB}(\theta)$ with two resources e_1, e_2. Assume without loss of generality that the total amount of players is unitary. We have that $\ell_1(x) = \alpha_1 x + \beta_1$ and $\ell_2(x) = \alpha_2 x + \beta_2$, for some $\alpha_1, \alpha_2, \beta_1, \beta_2 \geq 0$. We can resort to the primal-dual approach, but without necessarily passing trough the dual. Let x and y be the equilibrium and the optimal congestions of the first resource, respectively. We assume without loss of generality that $x \geq y$. To find an upper bound on the price of anarchy of $\mathsf{LB}(\theta)$, it is sufficient finding the maximum value of the following linear program $\mathsf{LP}(x,y)$ in variables $\alpha_1, \alpha_2, \beta_1, \beta_2, \geq 0$.

$$\begin{aligned}
\max \quad & \alpha_1 x^2 + \beta_1 x + \alpha_2(1-x)^2 + \beta_2(1-x) \\
s.t. \quad & \alpha_1 x + \beta_1 \leq \alpha_2(1-x) + \beta_2 \\
& \beta_2 \leq \theta\beta_1 \\
& \alpha_1 y^2 + \beta_1 y + \alpha_2(1-y)^2 + \beta_2(1-y) = 1,
\end{aligned}$$

where, as usual, the objective function coincides with the social cost of the equilibrium, the first constraint models some relaxed pure Nash equilibrium conditions, the second constraint imposes some relaxed θ-similarity conditions, and the third constraint normalizes the optimal social cost.

By standard arguments, one can show that the maximum value of $\mathsf{LP}(x,y)$ is attained by setting $\alpha_2 := 0$ and by imposing the tightness of all the constraints; under this choice, we have that $\alpha_1 x = (\theta-1)\beta_1$, $\beta_2 = \theta\beta_1$, and β_1 is such that the normalization constraint holds. By using these facts, and by simple calculations, we get that the optimal value is equal to

$$\frac{\alpha_1 x^2+\beta_1 x+\alpha_2(1-x)^2+\beta_2(1-x)}{\alpha_1 y^2+\beta_1 y+\alpha_2(1-y)^2+\beta_2(1-y)} = \frac{(\theta-1)x^2-(\theta-1)x+\theta}{(\theta-1)y^2/x+y+\theta(1-y)}. \tag{6.11}$$

By standard computations, one can easily show that (6.11) is maximized by $x=1$. Then, by further optimizing over y, we get that $y := \frac{1}{2}$ maximizes (6.11). Then, by using these values of x and y in (6.11), we get that the maximum value of (6.11) is $\frac{4\theta}{3\theta+1}$, and by the above considerations, this is an upper bound on $\mathsf{PoA}(\mathsf{LB})$. □

By reversing the proof arguments of Theorem 6.3, we can easily provide a tight lower bound.

Theorem 6.4. *There exists an arbitrary symmetric load balancing game* $\mathsf{LB}(\theta)$ *with two resources, affine latencies and* θ*-similar strategies such that* $\mathsf{PoA}(\mathsf{LB}) = \frac{4\theta}{3\theta+1}$.

Proof. It is sufficient considering a game LB as in the proof of Theorem 6.3, where the coefficients $\alpha_1,\alpha_2,\beta_1,\beta_2$, after a proper rescaling, are such that the linear program $\mathsf{LP}(x,y)$ is maximized, where x and y are set equal to 1 and $1/2$, respectively, i.e., we can set $\alpha_1 := (\theta-1)\beta_1$, $\alpha_2 := 0$, $\beta_2 := \theta\beta_1$, and $\beta_1 := 1$ in the latencies of the considered game. As $\beta_2 := \theta\beta_1$, we have that the game is with θ-similar strategies. By reusing the proof steps of Theorem 6.3, one can easily show that $x=1$ and $y=1/2$ are respectively the equilibrium and the optimal congestion of the first resource, and this turns into a price of anarchy of at least $\frac{4\theta}{3\theta+1}$. □

6.4 Concluding Remarks and Related Work

This chapter is based on some results obtained by Bilò and Vinci [29], who introduce the new model of congestion games with θ-similar strategies, and study their price of anarchy for both atomic and non-atomic affine congestion games. The results on non-atomic games have been improved and elegantly extended to the case of general latency functions by Benita et al. [18]. In particular, they consider the equivalent model of θ-*free-flow games*, and provide quasi-explicit formulas for computing the price of anarchy of general θ-free-flow non-atomic congestion games with general latency functions, which are tight even for non-symmetric load balancing games. Furthermore, they provide quasi-explicit formulas for computing the exact price of anarchy of parallel-link networks (i.e., symmetric load balancing games). As a byproduct of their analysis, they characterize the worst-case price of anarchy of classical non-atomic congestion games with *homogeneous latencies* (i.e., containing latencies ℓ with $\ell(0) = 0$ only), and show that it is already attained by parallel-link networks. This has solved a long-standing open problem posed by Roughgarden [155]. Furthermore, in [18] the authors also studied the road traffic in Singapore by analysing some data, with the further aim of validating the effectiveness of θ-free-flow games as new model for selfish routing.

Chapter 7
Non-Atomic One-Round Walks

In this chapter, we translate the solution concept of one-round walks from atomic congestion games to non-atomic ones, in order to describe what happens to the performance of one-round walks when the contribution of each player to the social cost is infinitesimally small. To this aim, we define the solution concept of ε*-approximate non-atomic one-round walk starting from the empty state*, in which there is a continuous flow of $R > 0$ selfish players (instead of a discrete number of players) greedily selecting their strategies with the aim of approximately minimizing their costs (up to a factor of $1+\varepsilon$), given the choices of their predecessors. In particular, we formally define a non-atomic one-round walk as a family $(\boldsymbol{\sigma}^t)_{t\in[0,R]}$ of (infinite) strategy profiles such that each strategy profile $\boldsymbol{\sigma}^t$ is a configuration obtained when an amount of t players (according to a given arrival ordering) has already entered the game, and each player $s \in [0,t]$ has chosen irrevocably her best-response, with respect to the strategy profile $\boldsymbol{\sigma}^s$ (i.e., the partial configuration realised at her arrival time). This is similar to what happens when considering one-round walks in atomic games, in which the players enter the game sequentially and select their best-response, but in discrete-time steps. Furthermore, one can observe that the concept of non-atomic one-round walk "approximates" in some way the behavior of its atomic counterpart, so as a differential (or integral) equation can be used to approximate the behavior of some equations recursively defined.

To measure the performance of non-atomic one-round walks, analogously to the price of anarchy for the solution concept of Nash equilibrium, we consider the *competitive ratio* of ε-approximate non-atomic one-round walks starting from the empty state, that is the highest ratio between the social value achieved by the final outcome of an ε-approximate non-atomic one-round walk and the optimal social value.

We derive an explicit formula to determine the ε-approximation ratio for congestion games having general latency functions. Furthermore, we study this metric for non-atomic congestion games with polynomial latencies, by proving a tight bound of $((1+\varepsilon)(d+1))^{d+1}$, where d is the degree of the polynomial. Given that the (exact and approximate) price of anarchy for these games has been shown to be equal to $\Theta(d/\log(d))$ (Roughgarden [155]), this result shows that outcomes generated

V. Bilò, C. Vinci, *Coping with Selfishness in Congestion Games*, Monographs in Theoretical Computer Science. An EATCS Series, https://doi.org/10.1007/978-3-031-30261-9_7

after one-round walks are worse (even asymptotically) than pure Nash equilibria in terms of social welfare.

7.1 Model and Definitions

We recall the model of non-atomic congestion games from Chapter 2 (in particular, from Section 2.5). Let

$$\mathsf{NCG} = (N, E, (\ell_e)_{e\in E}, (r_i)_{i\in N}, (\Sigma_i)_{i\in N})$$

be an arbitrary non-atomic game. Let $R := \sum_{i\in N} r_i$ denote the total amount of flow of the game. A function $H : [0,R] \to N$ such that $H^{-1}[\{i\}]$ is the union of intervals of type $[a,b[$ or $[a,R]$, and verifies $\int_{H^{-1}[\{i\}]} ds = r_i$ for each $i \in N$, is called *ordering function*.

Definition 7.1. Let $\boldsymbol{\tau} := [(\boldsymbol{\sigma}^t)_{t\in[0,R]}, H]$ be a family of strategy profiles (with $\boldsymbol{\sigma}^t = (\sigma^t_{i,S})_{i\in N, S\in\Sigma_i}$) equipped with an ordering function H satisfying the following conditions:

Condition 1: $\sum_{S\in\Sigma_i} \sigma^t_{i,S} = \int_{[0,t]\cap H^{-1}[\{i\}]} ds$ for each $i \in N$ and $t \in [0,R]$;
Condition 2: $\sigma^t_{i,S}$ is non-decreasing in t, for any $i \in N$ and $S \in \Sigma_i$.

Such $\boldsymbol{\tau}$ is called *weak (non-atomic) one-round walk starting from the empty state*.

Observe that $\sigma^t_{i,S}$, as a function from $t \in [0,R]$ to $\mathbb{R}_{\geq 0}$, has to be necessarily non-decreasing and Lipschitz continuous with respect to t; furthermore, its right-derivative, denoted as $\frac{\partial^+ \sigma^t_{i,S}}{\partial t}$, is well-defined and non-negative for any $t \in [0,R]$, $i \in N$, $S \in \Sigma_i$, and

$$\sum_{i\in N}\sum_{S\in\Sigma_i} \frac{\partial^+ \sigma^t_{i,S}}{\partial t} = 1, \ \forall t \in [0,R]. \tag{7.1}$$

Informally speaking, a weak one-round walk models a family of strategy profiles generated by a flow of players sequentially selecting their strategies, in such a way that, for any $t \in [0,R]$, an amount t of players has entered the game and, for any $i \in N$ and strategy S, there is an amount $\sigma^t_{i,S}$ of players of type i which have already selected strategy S (Condition 1 of Definition 7.1), and these players cannot change their strategy in subsequent steps (Condition 2 of Definition 7.1). Moreover, observe that the underlying ordering function H defines the ordering in which the players appear in the game. The strategy profile generated by a weak one-round walk is $\boldsymbol{\sigma}^R$, that is the strategy profile obtained when all players have entered the game.

Definition 7.2. An *ε-approximate (non-atomic) one-round walk starting from the empty state involving selfish players* (or *ε-one-round walk*, for brevity) is a weak one-round walk $\boldsymbol{\tau}$ such that, for any $t \in [0,R]$, $\sigma^t_{H(t),S}$ is right-increasing at t only if strategy S minimizes, up to a factor $1+\varepsilon$, the cost of player $H(t)$ in the partial strategy profile $\boldsymbol{\sigma}^t$.

Informally, an ε-one-round walk is a weak one-round walk in which, for any $t \in [0,R]$, the player entering the game at time t (i.e., of type $H(t)$) selects an ε-approximate best response with respect the partial strategy profile $\boldsymbol{\sigma}^t$. Given a congestion game CG, let $\mathsf{ORW}^s_\varepsilon(\mathsf{CG})$ be the set of strategy profiles generated by some ε-approximate one-round walk in CG. The concept of competitive ratio is defined as in Section 2.4, but applied to the concept of non-atomic one-round walks defined here. We denote the competitive of ε-one-round walks of a game CG (resp. class $\mathcal{G}$) by $\mathsf{CR}^s_\varepsilon(\mathsf{CG})$ (resp. $\mathsf{CR}^s_\varepsilon(\mathcal{G})$).

To illustrate the main definitions given in this section, we provide a clarifying example.

Example 7.1. Consider a load balancing game CG with three resources e_1, e_2, e_3, having latency functions $f_1(x) = x$, $f_2(x) = 2x$ and $f_3(x) = 4$ respectively, and with two types of players, such that $r_1 = 3$, $r_2 = 4$, $\Sigma_1 = \{\{e_1\},\{e_2\}\}$ and $\Sigma_2 = \{\{e_2\},\{e_3\}\}$. Consider the ordering function H such that $H(t) = 1$ if $t < 3$ and $H(t) = 2$ if $t \in [3,7]$. Let $\boldsymbol{\tau} := [(\boldsymbol{\sigma}^t)_{t\in[0,R]}, H]$ be the weak one-round walk such that

- given $t \in [0,3)$, we have $\sigma^t_{1,\{e_1\}} = 2t/3$, $\sigma^t_{1,\{e_2\}} = t/3$, and $\sigma^t_{2,\{e\}} = 0$ for any $e \in \{e_2, e_3\}$;
- $\sigma^t_{1,\{e_1\}} = 2$, $\sigma^t_{1,\{e_2\}} = 1$, $\sigma^t_{2,\{e_2\}} = t - 3$ and $\sigma^t_{2,\{e_3\}} = 0$, if $t \in [3,4)$;
- $\sigma^t_{1,\{e_1\}} = 2$, $\sigma^t_{1,\{e_2\}} = 1$, $\sigma^t_{2,\{e_2\}} = 1$ and $\sigma^t_{2,\{e_3\}} = t - 4$, if $t \in [4,7)$.

One can observe that $\boldsymbol{\tau}$ is a 0-one-round walk, and that the resulting strategy profile $\boldsymbol{\sigma} := \boldsymbol{\sigma}^R$ is such that $k_1(\boldsymbol{\sigma}) = k_2(\boldsymbol{\sigma}) = 2$ and $k_3(\boldsymbol{\sigma}) = 3$.

7.2 Competitive Ratio

In this section, by using the primal-dual method in the same spirit as in Theorem 4.1 from Chapter 4, we give a quasi-explicit formula to compute a tight bound on $\mathsf{CR}^s_\varepsilon(\mathsf{N}(\mathcal{C}))$. We first give some preliminary definitions. Given $x \geq 1$, $k > 0$, $o > 0$, and a latency function f, let

$$\gamma_{\mathsf{N}}(CR^s_\varepsilon, k, o, f) := \frac{kf(k) + x \cdot \beta_{\mathsf{N}}(\mathsf{CR}^s_\varepsilon, k, o, f)}{of(o)}, \tag{7.2}$$

where

$$\beta_{\mathsf{N}}(\mathsf{CR}^s_\varepsilon, k, o, f) := -\int_0^k f(s)\mathrm{d}s + (1+\varepsilon)of(k); \tag{7.3}$$

furthermore, given a class of non-atomic congestion games $\mathcal{G}$, let

$$\gamma_{\mathsf{N}}(\mathsf{CR}^s_\varepsilon, \mathcal{G}) := \inf_{x\geq 1} \sup_{f\in\mathcal{C}(\mathcal{G}),k,o>0} \gamma_{\mathsf{N}}(\mathsf{CR}^s_\varepsilon, x, k, o, f). \tag{7.4}$$

7.2.1 Upper Bound

We first show that (7.4) is an upper bound on the competitive ratio.

Theorem 7.1. *For any class of non-atomic congestion games* $\mathcal{G}$*, we have that* $\mathsf{CR}^s_\varepsilon(\mathcal{G}) \leq \gamma_{\mathsf{N}}(\mathsf{CR}^s_\varepsilon, \mathcal{G})$.

Proof. Let $\mathsf{NCG} \in \mathsf{N}(\mathcal{C})$ be a non-atomic congestion game. Let $\boldsymbol{\tau} = [(\boldsymbol{\sigma}^t)_{t\in[0,R]}, H]$ be an ε-one-round walk generating strategy profile $\boldsymbol{\sigma} := \boldsymbol{\sigma}^R$, and let $\boldsymbol{\sigma}^*$ be the optimal strategy profile of NCG. Let $\boldsymbol{\tau}^*$ be a weak one-round walk generating strategy profile $\boldsymbol{\sigma}^*$, based on the same ordering function H of $\boldsymbol{\tau}$, and such that, for any $t \in [0,R]$, there exists a (unique) strategy $S^*(t)$ such that $\frac{\partial^+ \sigma^{*,t}_{H(t),S^*(t)}}{\partial t} = 1$ (so that $\frac{\partial^+ \sigma^{*,t}_{i,S}}{\partial t} = 0$ if $i \neq H(t)$ or $S \neq S^*(t)$). We observe that such a weak one-round walk do exist.[1]

As usual, we set $k_e := k_e(\boldsymbol{\sigma})$ and $o_e := k_e(\boldsymbol{\sigma}^*)$. We have that the maximum value of the following linear program (with infinitely many constraints parametrized by $t \in [0,R]$) in the variables α_es is an upper bound on $\mathsf{CR}^s_\varepsilon(\mathsf{NCG})$:

$$\begin{aligned}
&\mathsf{LP1}: \\
&\max \quad \sum_{e\in E} \alpha_e k_e \ell_e(k_e) \\
&s.t. \quad \sum_{e\in S} \alpha_e \ell_e\left(k_e\left(\boldsymbol{\sigma}^t\right)\right) \leq (1+\varepsilon) \sum_{e\in S^*(t)} \alpha_e \ell_e\left(k_e\left(\boldsymbol{\sigma}^t\right)\right), \\
&\qquad\qquad \forall t \in [0,R],\ S \in \Sigma_{H(t),\varepsilon}(\boldsymbol{\sigma}^t) \qquad (7.5) \\
&\qquad \sum_{e\in E} \alpha_e o_e \ell_e(o_e) = 1 \qquad (7.6) \\
&\qquad \alpha_e \geq 0, \quad \forall e \in E,
\end{aligned}$$

where $\Sigma_{H(t),\varepsilon}(\boldsymbol{\sigma}^t)$ denotes the set of strategies S of player $H(t)$ such that $\sigma^t_{H(t),S}$ is right-increasing in t, and $S^*(t)$ denotes the strategy of player $H(t)$ such that $\frac{\partial^+ \sigma^{*,t}_{H(t),S^*(t)}}{\partial t} = 1$. Indeed, by setting $\alpha_e = 1$ for any $e \in E$, we have that:

- the objective function of $\mathsf{LP1}$ is the total latency $\mathsf{SUM}(\boldsymbol{\sigma})$;
- (7.5) holds since $[(\boldsymbol{\sigma}^t)_{t\in[0,R]}, H]$ is an ε-one-round walk, thus any strategy S for which $\sigma^t_{H(t),S}$ is right-increasing in t is necessarily an approximate best-response for players of type $H(t)$;
- (7.6) is the normalized optimal total latency $\mathsf{SUM}(\boldsymbol{\sigma}^*)$.

[1] To show the existence of such a weak one-round walk, it is sufficient considering the following continuous-time process. At each time $t \in [0,R)$, consider a strategy $S^* \in \Sigma_{H(t)}$ such that $\sigma^{*,t}_{H(t),S^*} < \sigma^*_{H(t),S^*}$. Then, assign players of type $H(t)$ to this strategy only, until either $\sigma^{*,t'}_{H(t'),S^*} = \sigma^*_{H(t'),S^*}$ holds, or $H(t) \neq H(t')$, for some subsequent step $t' > t$. Repeat this process with respect to the new step t^*, and iterate it until the total amount of players R has been assigned.

Now, we introduce some relaxations on LP1 to get a new linear program. As $k_e(\boldsymbol{\sigma}^t)$ is non-decreasing in t, we have that $k_e(\boldsymbol{\sigma}^t) \leq k_e$, so that we can replace $(1+\varepsilon)\sum_{e\in S^*(t)} \alpha_e f_e(k_e(\boldsymbol{\sigma}^t))$ with $(1+\varepsilon)\sum_{e\in S^*(t)} \alpha_e f_e(k_e)$ in (7.5) to obtain a new relaxed constraint. Then, we can multiply each obtained constraint by the right-derivative $\frac{\partial^+ \sigma_{i,S}^t}{\partial t}$. Furthermore, if $S \notin \Sigma_{H(t),\varepsilon}(\boldsymbol{\sigma}^t)$ or $i \neq H(t)$, we have that $\sigma_{i,S}^t$ is constant in a right-neighborhood of t, and then $\frac{\partial^+}{\partial t}\sigma_{i,S}^t = 0$. Hence, we can extend the new relaxed constraints constructed from (7.5) to strategies $S \notin \Sigma_{H(t),\varepsilon}(\boldsymbol{\sigma}^t)$ and types $i \neq H(t)$. Thus, the final relaxed constraint that we can adopt in place of (7.5) is the following:

$$\frac{\partial^+ \sigma_{i,S}^t}{\partial t}\sum_{e\in S}\alpha_e \ell_e\left(k_e\left(\boldsymbol{\sigma}^t\right)\right) \leq \frac{\partial^+ \sigma_{i,S}^t}{\partial t}(1+\varepsilon)\sum_{e\in S^*(t)}\alpha_e \ell_e(k_e), \ \forall i\in N,\ S\in\Sigma_i. \quad (7.7)$$

To obtain a further relaxation of LP1, we can integrate each constraint of type (4.4) with respect to variable $t \in [0,R]$, and then we can sum the integrated constraints over $i \in N$ and $S \in \Sigma_i$. Thus, we get the following inequalities:

$$\begin{aligned}
0 &\geq \sum_{i\in N}\sum_{S\in\Sigma_i}\int_0^R \left(\sum_{e\in S}\alpha_e\ell_e\left(k_e\left(\boldsymbol{\sigma}^t\right)\right) - (1+\varepsilon)\sum_{e\in S^*(t)}\alpha_e\ell_e(k_e)\right)\frac{\partial^+\sigma_{i,S}^t}{\partial t}dt\\
&= \sum_{e\in E}\alpha_e\int_0^R \ell_e\left(k_e(\boldsymbol{\sigma}^t)\right)\sum_{i\in N, S\in\Sigma_i : e\in S}\frac{\partial^+\sigma_{i,S}^t}{\partial t}dt\\
&\quad -(1+\varepsilon)\int_0^R \sum_{e\in S^*(t)}\alpha_e\ell_e(k_e)\overbrace{\sum_{i\in N}\sum_{S\in\Sigma_i}\frac{\partial^+\sigma_{i,S}^t}{\partial t}}^{=1}dt\\
&= \sum_{e\in E}\alpha_e\int_0^{k_e}\ell_e\left(k_e(\boldsymbol{\sigma}^t)\right)\mathrm{d}\left(\overbrace{\sum_{i\in N, S\in\Sigma : e\in S}\sigma_{i,S}^t}^{=k_e(\boldsymbol{\sigma}^t)}\right)\\
&\quad -(1+\varepsilon)\int_0^R\sum_{e\in S^*(t)}\alpha_e\ell_e(k_e)dt\\
&= \sum_{e\in E}\alpha_e\int_0^{k_e}\ell_e\left(k_e(\boldsymbol{\sigma}^t)\right)\mathrm{d}(k_e(\boldsymbol{\sigma}^t))\\
&\quad -(1+\varepsilon)\int_0^R\sum_{i\in N}\sum_{S^*\in\Sigma_i}\frac{\partial^+\sigma_{i,S^*}^{*,t}}{\partial t}\sum_{e\in S^*}\alpha_e\ell_e(k_e)dt\\
&= \sum_{e\in E}\alpha_e\int_0^{k_e}\ell_e(s)\,\mathrm{d}s\\
&\quad -(1+\varepsilon)\sum_{e\in E}\alpha_e\left(\sum_{i\in N, S^*\in\Sigma_i : e\in S^*}\int_0^R\frac{\partial^+\sigma_{i,S^*}^{*,t}}{\partial t}dt\right)\ell_e(k_e)
\end{aligned}$$

$$= \sum_{e\in E} \alpha_e \int_0^{k_e} \ell_e(s)\,ds - (1+\varepsilon)\sum_{e\in E}\alpha_e \Bigg(\overbrace{\sum_{i\in N, S^*\in\Sigma_i : e\in S^*} \sigma^*_{i,S^*}}^{=o_e} \Bigg) \ell_e(k_e)$$

$$= \sum_{e\in E} \alpha_e \int_0^{k_e} \ell_e(s)\,ds - (1+\varepsilon)\sum_{e\in E} \alpha_e o_e \ell_e(k_e). \tag{7.8}$$

We conclude that, by replacing (7.5) with (7.8), we obtain the following relaxation of $\mathsf{LP1}$:

$$\begin{aligned} \mathsf{LP2}: \quad \max \quad & \sum_{e\in E} \alpha_e k_e \ell_e(k_e) \\ s.t. \quad & \sum_{e\in E} \alpha_e \int_0^{k_e} \ell_e(s)ds \le (1+\varepsilon)\sum_{e\in E} \alpha_e o_e \ell_e(k_e) \\ & \sum_{e\in E} \alpha_e o_e \ell_e(o_e) = 1 \\ & \alpha_e \ge 0, \quad \forall e\in E. \end{aligned}$$

Now, by proceeding as in Theorem 4.1, we compute the dual of $\mathsf{LP2}$ (with two variables x,γ), that has the following form:

$$\begin{aligned} \mathsf{DLP}: \quad \min \quad & \gamma \\ s.t. \quad & \gamma \cdot o_e \ell_e(o_e) \ge \frac{kf(k) + x\cdot \beta_{\mathsf{N}}(\mathsf{CR}^s_\varepsilon, k_e, o_e, \ell_e)}{o_e \ell_e(o_e)}, \ \forall e\in E \\ & x \ge 0. \end{aligned}$$

Analogously to Theorem 4.1, one can show that $\gamma_{\mathsf{N}}(\mathsf{CR}^s_\varepsilon, \mathcal{G})$ is an upper bound on the optimal solution of DLP, that is an upper bound on $\mathsf{CR}^s_\varepsilon(\mathcal{G})$ (by the Weak Duality Theorem). □

7.2.2 Tight Lower Bound

In this subsection, we show that the upper bound provided in Theorem 7.1 is tight. We first give a preliminary lemma, whose claim and proof are similar to those of Lemma 4.1 of Chapter 4.

Lemma 7.1. *Let $\mathcal{G}$ be a class of non-atomic congestion games. For each $M < \gamma_{\mathsf{W}}(\mathsf{CR}^s_\varepsilon, \mathcal{G})$ one of the following cases holds:*

- **Case 1:** *there exist a latency function $f\in\mathcal{C}(\mathcal{G})$ and two rational numbers $k,o > 0$ such that*

$$M < \frac{kf(k)}{of(o)}$$

and $\beta_{\mathsf{N}}(\mathsf{CR}^s_\varepsilon, k, o, f) \geq 0$.

- **Case 2:** *there exist two latency functions* $f_1, f_2 \in \mathcal{C}(\mathcal{G})$ *and four rational numbers* $k_1, k_2, o_1, o_2 > 0$ *such that*

$$M < \frac{\alpha_1 k_1 f_1(k_1) + \alpha_2 k_2 f_2(k_2)}{\alpha_1 o_1 f_1(o_1) + \alpha_2 o_2 f_2(o_2)},$$

where $\alpha_1 := \beta_{\mathsf{N}}(\mathsf{CR}^s_\varepsilon, k_2, o_2, f_2) > 0$ *and* $\alpha_2 := -\beta_{\mathsf{N}}(\mathsf{CR}^s_\varepsilon, k_1, o_1, f_1) > 0$.

By using Lemma 7.1 and by resorting to a similar result as in Theorem 4.8, for any fixed $M < \gamma_{\mathsf{N}}(\mathsf{CR}^s_\varepsilon, \mathcal{G})$, one can provide a class of non-atomic load balancing games $\mathsf{NCG} := \mathsf{NCG}_s(k_1, k_2, o_1, o_2, f_1, f_2)$, parametrized by the parameters $k_1, k_2, o_1, o_2, f_1, f_2$ of Lemma 7.1 and an integer $s \geq 1$, such that, for a sufficiently large s, $\mathsf{CR}^s_\varepsilon(\mathsf{NCG}) > M$ holds. This shows that $\gamma_{\mathsf{W}}(\mathsf{CR}^s_\varepsilon, \mathcal{G})$ is a tight lower bound even for load balancing games, if the considered latency functions are closed under ordinate-scaling.

Theorem 7.2. *We have that* $\gamma_{\mathsf{N}}(\mathsf{CR}^s_\varepsilon, \mathsf{N}(\mathcal{C})) = \mathsf{CR}^s_\varepsilon(\mathsf{N}(\mathcal{C})) = \mathsf{CR}^s_\varepsilon(\mathsf{NLB}(\mathcal{C}))$ *for any class* $\mathcal{C}$ *of latency functions.*

In the following theorem we show a weaker statement, and we provide a general non-atomic congestion game as a tight lower bound.

Theorem 7.3. *We have that* $\gamma_{\mathsf{N}}(\mathsf{CR}^s_\varepsilon, \mathsf{N}(\mathcal{C})) = \mathsf{CR}^s_\varepsilon(\mathsf{N}(\mathcal{C}))$ *for any class* $\mathcal{C}$ *of latency functions.*

Proof (Sketch). For the sake of simplicity, we provide the main intuitions on how to construct the tight lower bound matching the competitive ratio of Theorem 7.1. Instead of providing a non-atomic game as a lower bound, we provide an unweighted atomic game $\mathsf{CG}_{t,n}$ as that defined in Theorem 4.10, in which the number of players selecting each resource is proportional to an integer parameter $t \geq 1$, and the contribution that each player gives to the congestion of each resource is proportional to $1/t$. At the end of the proof, one can easily see that, for sufficiently large values of t, the competitive ratio of $\mathsf{CG}_{t,n}$ well approximates (with arbitrary precision) that of a non-atomic congestion game $\mathsf{NCG}_{t,n}$ obtained from $\mathsf{CG}_{t,n}$, by replacing each atomic player with an amount of non-atomic ones and by adding further dummy resources that preserve the players' strategic behavior.

Fix $M < \gamma_{\mathsf{W}}(\mathsf{CR}^s_\varepsilon, \mathcal{G})$. Let $k_1, k_2, o_1, o_2, f_1, f_2$ be the parameters provided by Lemma 7.1, where we set $k_1 = k_2 = k$, $o_1 = o_2 = o$, and $f_1 = f_2 = f$ if the first case of the lemma holds. As k_1, k_2, o_1, o_2 are rational (by the hypothesis), by multiplying such numbers by an opportune integer $\hat{t}$ they become integers. For any integer t, let $\mathsf{CG}_{t,n} := \mathsf{CG}_n(k_{1,t}, k_{2,t}, o_{1,t}, o_{1,t}, f_{1,t}, f_{1,t})$ be the parametric lower bounding instance defined in Theorem 4.10, where $n := n_t$ is a sufficiently large integer, $k_{j,t} := k_j \cdot \hat{t} \cdot t$, $o_{j,t} := o_j \cdot \hat{t} \cdot t$, and $f_{t,j}(x) := f_f(x/(\hat{t} \cdot t))$ for any $j \in [2]$ and $x \geq 0$. Let $\boldsymbol{\sigma}_{t,n}$ be the strategy profile generated by the ε-one-round walk in which each player i selects strategy s_i, and let $\boldsymbol{\sigma}^*_{t,n}$ be the strategy profile in which each player i selects strategy s^*_i, except for those players i who can choose s_i only.

By using a similar approach as in the proof of Theorem 4.10, one can show that

$$\limsup_{t\to\infty}\limsup_{n\to\infty}\frac{\mathsf{SUM}(\boldsymbol{\sigma}_{t,n})}{\mathsf{SUM}(\boldsymbol{\sigma}^*_{t,n})}=\frac{\alpha_1 k_1 f_1(k_1)+\alpha_2 k_2 f_2(k_2)}{\alpha_1 o_1 f_1(o_1)+\alpha_2 o_2 f_2(o_2)}>M,$$

where $\alpha_1 := \beta_{\mathsf{N}}(\mathsf{CR}^s_\varepsilon, k_2, o_2, f_2) > 0$ and $\alpha_2 := -\beta_{\mathsf{N}}(\mathsf{CR}^s_\varepsilon, k_1, o_1, f_1) > 0$ if the second case of Lemma 7.1 holds, and $\alpha_1 := \alpha_2 := 1$ otherwise.[2] Thus, by taking sufficiently large t and n, we get $\mathsf{CR}^s_\varepsilon(\mathsf{CG}_{t,n}) > M$.

Finally, as observed at the beginning of the proof, if t and n are sufficiently large, $\mathsf{NCG}_{t,n}$ can be transformed into a non-atomic instance $\mathsf{NCG}_{t,n}$ guaranteeing a competitive ratio approximately equal to that of $\mathsf{CG}_{t,n}$ (with arbitrary precision), so that the resulting competitive ratio stays above M. By the arbitrariness of M, the claim follows. □

7.2.3 Application to Polynomial Latency Functions

By using the fact that $\gamma_{\mathsf{N}}(\mathsf{CR}^s_\varepsilon, \mathsf{N}(\mathcal{C}))$ is a tight bound on the competitive ratio, even for load balancing instances if $\mathcal{C}$ is closed under ordinate-scaling, it is sufficient computing the exact value of $\gamma_{\mathsf{N}}(\mathsf{CR}^s_\varepsilon, \mathsf{N}(\mathcal{P}(d)))$ to provide tight bounds for the restricted case of polynomial latency functions of maximum degree d.

Theorem 7.4. *The competitive ratio of ε-one-round walks for the class of non-atomic congestion games with polynomial latency functions of maximum degree $d \geq 1$ is $((1+\varepsilon)(d+1))^{d+1}$. This bound is tight even for load balancing games.*

Proof. By standard calculations, one can show that

$$\begin{aligned}
&\gamma_{\mathsf{N}}(\mathsf{CR}^s_\varepsilon, \mathsf{N}(\mathcal{P}(d)))\\
&=\inf_{x\geq 1}\sup_{f\in\mathcal{P}(d),k,o>0}\frac{kf(k)+x\left(-\int_0^k f(s)\mathrm{d}s+(1+\varepsilon)of(k)\right)}{of(o)}\\
&=\inf_{x\geq 1}\sup_{\alpha_1,\ldots,\alpha_d,k,o>0}\frac{k\sum_{h=0}^d \alpha_h k^h+x\left(-\int_0^k\left(\sum_{h=0}^d\alpha_h s^h\right)\mathrm{d}s+(1+\varepsilon)o\sum_{h=0}^d\alpha_h k^h\right)}{o\sum_{h=0}^d\alpha_h o^h}\\
&=\inf_{x\geq 1}\sup_{\alpha_1,\ldots,\alpha_d,k,o>0}\frac{\sum_{h=0}^d\alpha_h\left[k^{h+1}+x\left(-\frac{k^{h+1}}{h+1}+(1+\varepsilon)ok^h\right)\right]}{\sum_{h=0}^d\alpha_h o^{h+1}}
\end{aligned}$$

[2] We observe that the limit process for t tending to infinite has transformed the sums of type $\sum_{h=1}^k f(h)$ appearing in the competitive ratio for the atomic case, into integrals of type $\int_0^k f(s)ds$, which instead appear in the quantities $\beta_{\mathsf{N}}(\mathsf{CR}^s_\varepsilon, k_2, o_2, f_2)$ and $-\beta_{\mathsf{N}}(\mathsf{CR}^s_\varepsilon, k_1, o_1, f_1)$ defined here.

$$= \inf_{x\geq 1} \sup_{h\in[d]_0,k,o>0} \frac{k^{h+1} + x\left(-\frac{h^{d+1}}{h+1} + (1+\varepsilon)ok^h\right)}{o^{h+1}}$$

$$= \inf_{x\geq 1} \sup_{h\in[d]_0,t>0} \left[t^{h+1} + x\left(-\frac{t^{h+1}}{h+1} + (1+\varepsilon)ot^h\right)\right]$$

$$= \inf_{x\geq 1} \sup_{h\in[d]_0,t>0} \gamma(x,t,h),$$

where $\gamma(x,t,h) := t^{h+1} + x\left(-\frac{t^{h+1}}{h+1} + (1+\varepsilon)ot^h\right)$. Again by standard calculations, one can show that, for $x^* := (d+1)d$, $\gamma(x,t,h)$ is maximized by $t^* = (1+\varepsilon)(d+1)$ and $h = d$. Thus, we get that $\gamma_{\mathsf{N}}(\mathsf{CR}^s_\varepsilon, \mathsf{N}(\mathcal{P}(d)))$ is at most equal to $\gamma^* := \gamma(x^*,t^*,d) = ((d+1)(1+\varepsilon))^{d+1}$.

Now, we show that $\gamma_{\mathsf{N}}(\mathsf{CR}^s_\varepsilon, \mathsf{N}(\mathcal{P}(d)))$ is exactly equal to $\gamma^* = ((d+1)(1+\varepsilon))^{d+1}$. By similar arguments as in Remark 4.2, we have that $\frac{kf(k)}{of(o)}$ is a lower bound on $\gamma_{\mathsf{N}}(\mathsf{CR}^s_\varepsilon, \mathsf{N}(\mathcal{P}(d)))$, if $\beta_{\mathsf{N}}(\mathsf{CR}^s_\varepsilon, k, o, f) \geq 0$. We observe that, for $k := t^* = (1+\varepsilon)(d+1)$, $o := 1$, and f defined as $f(x) = x^d$, we get $\frac{kf(k)}{of(o)} = ((d+1)(1+\varepsilon))^{d+1}$ and $\beta_{\mathsf{N}}(\mathsf{CR}^s_\varepsilon, k, o, f) \geq 0$. Thus, we necessarily have that $((d+1)(1+\varepsilon))^{d+1} = \gamma_{\mathsf{N}}(\mathsf{CR}^s_\varepsilon, \mathsf{N}(\mathcal{P}(d)))$, and this shows the claim. □

7.3 Concluding Remarks and Related Work

The results presented in this chapter are based on the works of Vinci [174, 175]. In particular, the results obtained in [174] are related to polynomial congestion games only, and the work [174] extends the previous results from polynomial latencies to general ones.

The framework considered in this chapter exhibits some similarities with an online optimization setting considered by Harks et al. [102]. The authors of [102] consider an online setting in which several fractions of the total players' flow arrive sequentially, and each fraction must be irrevocably split among the feasible strategies without knowing the future arrivals, so as to minimize the total latency when all the flow has been assigned. They focus on affine network congestion games and provide upper bounds on the competitive ratio guaranteed by the greedy algorithm, that splits each fraction of the flow so as to minimize the total latency; furthermore, they provide lower bounds on the best competitive ratio that can be achieved by any online algorithm.

Harks et al. [102] extended some of the above results to the case of polynomial latencies of maximum degree $d \geq 1$. In particular, they show that the competitive ratio of the greedy algorithm is at most $(d+1)^{d+1}$, and there are instances for which it is $\frac{d+1}{d+1}2^{d+1}$. Despite the setting presented in this chapter is different from that studied by Harks et al. [102], one can show that the lower bound instances considered in Theorem 7.2 and 7.3, after some minor modifications, can be reconsidered to show the tightness of the upper bound $(d+1)^{d+1}$ on the competitive ratio of the greedy algorithm, thus improving on the previous lower bound of Harks et al. [102].

Part IV
Taxes and Other Strategies to Cope with Selfish Behavior

Chapter 8
Taxes for Congestion Games: Preliminaries

In Part III, we have focused on the performance of congestion games under different efficiency metrics (price of anarchy, price of stability, competitive ratio), and the presented results have shown that selfish behavior produces suboptimal outcomes, whose performance may be quite bad in certain cases. Thus, it becomes natural to ask whether or nor the impact of selfish behavior on the system efficiency can be mitigated by an external mild intervention. Toward this end, the use of taxation (Caragiannis et al. [52], Cole et al. [71], Pigou [150]) has been fruitfully proposed and analyzed in the literature of congestion games. This approach aims at discouraging the use of certain (usually highly congested) resources by forcing players to pay taxes for using them.

In this chapter, we provide the main definitions and the basic notation related to taxation mechanisms, and we give some mathematical background about combinatorics and number theory that we shall use in the subsequent chapters to define and analyze certain classes of taxes. We mainly focus on atomic congestion games; anyway, the considered definitions can be easily adapted to the non-atomic case.

In the last section, we provide references to the literature of taxation in congestion games and to some related settings. Several research works are not mentioned here, as they are discussed in some of the subsequent chapters which they are connected to.

8.1 Tax Functions

By following the setting of Caragiannis et al. [52], we consider the use of resource taxation in order to mitigate the impact of selfish behavior in weighted congestion games.

Definition 8.1 (dynamic tax function). A *dynamic tax function* is a function $\delta : E \times \mathbb{R}^2_{\geq 0} \to \mathbb{R}_{\geq 0}$ that assigns a tax $\delta_e(w, k_e)$ to each player of weight w that wishes to use resource $e \in E$ in strategy profile $\boldsymbol{\sigma}$.

V. Bilò, C. Vinci, *Coping with Selfishness in Congestion Games*, Monographs in Theoretical Computer Science. An EATCS Series, https://doi.org/10.1007/978-3-031-30261-9_8

By so doing, a congestion game CG extends to a new game (CG,δ) in which each player i wants to minimize a new cost function $\widehat{c}_i(\boldsymbol{\sigma}) = c_i(\boldsymbol{\sigma}) + \sum_{e\in\sigma_i}\delta_e(w_i,k_e)$. In such a setting, two new social functions are considered, so as to account also for the amount of taxation introduced in the game: the *total cost* $\mathsf{TC}(\boldsymbol{\sigma}) = \sum_{i\in N}\widehat{c}_i(\boldsymbol{\sigma})$ and the *weighted total cost* $\mathsf{WTC}(\boldsymbol{\sigma}) = \sum_{i\in N} w_i\widehat{c}_i(\boldsymbol{\sigma})$.

The tax function considered in Definition 8.1 is quite general (and powerful), as it may depend on both the global structure of the game and the realized strategy profile. The unique constraint to be satisfied is that taxes must be applied locally by each resource to each player selecting it, despite tax-functions may be defined globally. One can clearly consider subclasses of the tax functions defined in Definition 8.1, which could be less powerful, but simpler. This is reasonable in any setting in which a designer has a partial knowledge of the game or low computational power.

Definition 8.2. A dynamic tax function $\delta_e(w,k_e)$ is based on a mixed strategy profile $\mathbf{y}$ whenever its dependence on e is restricted to the structure of the latency function and the values $\mathbb{E}[k_e(\mathbf{y})]$ and $\mathbb{E}[n_e(\mathbf{y})]$ only, that is, $\delta_e(w,k_e) := \delta_{\ell_e}(w,\mathbb{E}[k_e(\mathbf{y})],\mathbb{E}[n_e(\mathbf{y})],k_e)$, and if, for any latency functions f_1, f_2, and real numbers $\alpha_1,\alpha_2 \geq 0$, the following linearity property holds:

$$\begin{aligned}&\delta_{\alpha_1 f_1+\alpha_2 f_2}(w,\mathbb{E}[k_e(\mathbf{y})],\mathbb{E}[n_e(\mathbf{y})],k_e)\\&=\alpha_1\delta_{f_1}(w,\mathbb{E}[k_e(\mathbf{y})],\mathbb{E}[n_e(\mathbf{y})],k_e)+\alpha_2\delta_{f_2}(w,\mathbb{E}[k_e(\mathbf{y})],\mathbb{E}[n_e(\mathbf{y})],k_e).\end{aligned}$$

A natural choice for $\mathbf{y}$, that we shall widely exploit in the subsequent chapters, is that of an optimal strategy profile.

Remark 8.1. Consider the efficiency of taxes for a given class $\mathcal{G}$ of congestion games. If all latency functions in $\mathcal{C}(\mathcal{G})$ can be defined as $\ell_e = \sum_{h=0}^{d}\alpha_{e,h}\ell_h$, where $\alpha_{e,h} \geq 0$ and $\{\ell_1,\ell_2,\ldots,\ell_d\}$ is a given set of latency functions, if we do not consider particular restrictions on the players' strategy space and if the considered taxes are based on an optimal strategy profile, we can also assume without loss of generality that such taxes satisfy the linearity property, without worsening the efficiency. Indeed, consider an arbitrary congestion game $\mathsf{CG} \in \mathcal{G}$ with latency functions of type $\ell_e = \sum_{h=0}^{d}\alpha_{e,h}\ell_h$, where we assume w.l.o.g. that each α_h is rational. We can construct an equivalent congestion game CG' by replacing each resource e of CG with an opportune finite number m_e of new resources, where each resource of the new game has latency functions belonging to $\{\ell_0,\ell_1,\ldots,\ell_d\}$, and the number of resources with latency ℓ_h is proportional to α_h. Let $\{\delta_0,\delta_1,\ldots,\delta_d\}$ be a set of taxes (based on a mixed strategy profile) associated to the set of latency functions $\{\ell_0,\ell_1,\ldots,\ell_d\}$. If we go back to the initial congestion game CG, we can assume that the tax function δ_e associated to each resource ℓ_e is equal to $\sum_{h=0}^{d}\alpha_{e,h}\delta_h$, and the efficiency of such taxes is the same as that guaranteed in CG' by taxes $\delta_0,\delta_1,\ldots,\delta_d$.

Example 8.1. An example of dynamic taxes based on a mixed strategy profile for congestion games with polynomial latency functions of maximum degree d is the following: $\delta_e(w,k_e) := \sum_{h\in[d]_0}\alpha_{e,h}f_h(w,\mathbb{E}[k_e(\mathbf{y})],\mathbb{E}[n_e(\mathbf{y})],k_e)$, where $\alpha_{e,h}$ is the coefficient of the monomial x^h in latency function ℓ_e (see Subsection 2.2), and for each $h\in[d]_0$, $f_h:\mathbb{R}^4_{\geq 0}\to\mathbb{R}_{\geq 0}$ is an arbitrary function.

Definition 8.3 (universal tax function). A dynamic tax function $\delta_e(w,k_e)$ is *universal* whenever its dependence on e is restricted to the structure of the latency function only, that is, $\delta_e(w,k_e) := \delta_{\ell_e}(w,k_e)$, and if the linearity property described in Definition 8.2 holds.

Definition 8.4 (static tax function). A *static tax function* is a dynamic tax function $\delta_e(w,k_e)$ that does not depend on the resource congestions in $\boldsymbol{\sigma}$, that is, $\delta_e(w,k_e) := \delta_e(w)$.

Observe that, by definition, a dynamic (resp. static) tax function is more powerful than a dynamic (resp. static) tax function based on a strategy profile, which is more powerful than a universal dynamic (resp. static) tax function; moreover, any dynamic tax function of a certain type is more powerful than any static tax function of the same type. We denote with $\mathsf{TAX}_{\mathsf{MSP}}(\mathcal{G})$ the set of all possible dynamic tax functions based on a mixed strategy profile that can be defined for the class of games $\mathcal{G}$ and with $\mathsf{TAX}_{\mathsf{OPT}}(\mathcal{G})$ the set of all possible dynamic tax functions based on an optimal strategy profile that can be defined for the class of games $\mathcal{G}$.

8.1.1 Efficiency of Taxation

As proposed by Cole et al. [71], two types of taxes may be considered: *refundable* and *non-refundable* ones. Refundable taxes are assumed to be returned to the players (for instance, as a lump-sum refund) and, for this reason, they do not contribute to the system disutility. Hence, the quality of a strategy profile in game (CG,δ) is measured by making use of the social functions TL and WTL. For non-refundable taxes, instead, social functions TC and WTC are used.

For a certain solution concept SC, let $\mathsf{SC}(\mathsf{CG},\delta)$ be the set of strategy profiles implementing SC in game (CG,δ).

Definition 8.5 (efficiency of refundable and non-refundable taxes). A tax function δ is a ρ-efficient refundable tax for the weighted congestion game CG with respect to the solution concept SC and to the social function $\mathsf{SF} \in \{\mathsf{TL},\mathsf{WTL}\}$, if $\mathsf{SF}(\boldsymbol{\sigma}) \leq \rho \cdot \mathsf{SF}(\boldsymbol{\sigma}^*)$ for each $\boldsymbol{\sigma} \in \mathsf{SC}(\mathsf{CG},\delta)$. A tax function δ is a ρ-efficient non-refundable tax for the weighted congestion game CG with respect to the solution concept SC and to the social function TC (resp. WTC), if $\mathsf{TC}(\boldsymbol{\sigma}) \leq \rho \cdot \mathsf{TL}(\boldsymbol{\sigma}^*)$ (resp. $\mathsf{WTC}(\boldsymbol{\sigma}) \leq \rho \cdot \mathsf{WTL}(\boldsymbol{\sigma}^*)$) for each $\boldsymbol{\sigma} \in \mathsf{SC}(\mathsf{CG},\delta)$.

We observe that the efficiency of a refundable (resp. non-refundable) tax function for a game CG is a concept analogue to price of anarchy, but the underlying solution concept SC is implemented by strategy profiles of the taxed game $\mathsf{SC}(\mathsf{CG},\delta)$ and the considered social function is $\mathsf{SF} \in \{\mathsf{TL},\mathsf{WTL}\}$ (resp. $\mathsf{SF} \in \{\mathsf{TC},\mathsf{WTC}\}$).

8.2 Mathematical Background on Combinatorics

In this subsection, we give an overview of the mathematical background on combinatorics and number theory, that we shall use in the following to determine and analyze the efficiency of some tax functions. For all the results claimed in this section, see Mansour and Schork [122].

Definition 8.6 (generalised Stirling numbers of the second kind). Given a pair of non-negative integers (i, j) and $\alpha \in \mathbb{R}$, the *generalised Stirling number of the second kind* $\mathcal{S}_\alpha(i, j)$ is defined by using the following recurrence:

$$\mathcal{S}_\alpha(0,0) = 1, \quad \mathcal{S}_\alpha(i,0) = \mathcal{S}_\alpha(0,i) = 0 \quad \forall i \in \mathbb{N}, \tag{8.1}$$

$$\mathcal{S}_\alpha(i+1, j) = (j+\alpha-1)\mathcal{S}_\alpha(i, j) + \mathcal{S}_\alpha(i, j-1). \tag{8.2}$$

By setting $\alpha = 1$ we obtain the well-known *Stirling numbers of the second kind*, simply denoted as $\mathcal{S}(i, j)$, which measure the number of different ways of partitioning a set of i elements into j non-empty subsets.

Example 8.2. $\mathcal{S}(3,2) = 3$ since the set $\{x,y,z\}$ can be partitioned into two subsets in three different ways as $\{\{x\},\{y,z\}\}$, $\{\{y\},\{x,z\}\}$ and $\{\{z\},\{x,y\}\}$.

Definition 8.7 (Touchard polynomials). For a positive integer i, the ith *Touchard* (or *Bell*) *polynomial* $\mathcal{B}_i(x)$ is defined as $\mathcal{B}_i(x) := \sum_{j=0}^{i} \mathcal{S}(i, j)x^j$.

By Dobiński's formula, we have

$$\mathcal{B}_i(x) = e^{-x} \sum_{p=0}^{\infty} \frac{p^i x^p}{p!}. \tag{8.3}$$

Definition 8.8 (geometric polynomials). The ith *Geometric (or Euler) polynomial* $\mathcal{E}_i(x)$ is defined as $\mathcal{E}_i(x) := \sum_{j=0}^{i} \mathcal{S}(i, j)\, j!x^j$.

By Euler's formula, we have

$$\mathcal{E}_i\left(\frac{x}{1-x}\right) = (1-x)\left(x\frac{\partial}{\partial x}\right)^i \frac{1}{1-x}, \tag{8.4}$$

where we recall that, given a function $f(x)$, $\left(x\frac{\partial}{\partial x}\right)^i f(x)$ is the function recursively defined as follows: $\left(x\frac{\partial}{\partial x}\right)^i f(x) = \left(x\frac{\partial}{\partial x}\right)^{i-1}\left(x\frac{\partial f(x)}{\partial x}\right)$ with $\left(x\frac{\partial}{\partial x}\right)^0 f(x) = f(x)$. It is immediate to see that, for any value $x \in \mathbb{R}$, both $\mathcal{B}_i(x)$ and $\mathcal{E}_i(x)$ are increasing functions of i.

Definition 8.9 (forward difference). The ith *forward difference* of function $f(x)$ is the function Δ^i recursively defined as follows: $\Delta^{i+1}[f(x)] = \Delta^i[f(x+1)] - \Delta^i[f(x)]$ with $\Delta^0[f(x)] = f(x)$.

It is known that $\Delta^i\left[x^h\right]$ is a polynomial of x with non-negative coefficients having degree equal to $h-i$, so that for each $x \in \mathbb{R}$, $\Delta^i\left[x^i\right] = \Delta^i\left[x^i\right]|_{x=0}$. Moreover, we have $\Delta^i\left[x^h\right]|_{x=0} = \mathcal{S}(h,i)i!$. Other polynomials we use are defined as follows:

$$\mathcal{Z}_i(x) := \sum_{j=0}^{i} j^{i-j}(i)_j x^j;$$

$$\mathcal{R}_i(x) := \sum_{j=0}^{i} \mathcal{S}_2(i,j)x^j;$$

$$\mathcal{Q}_i(x) := \sum_{j=0}^{i} x^j \sum_{h=j}^{i} i^j j^{i-h}\binom{i}{h}(-1)^{h+j}\mathcal{S}(h,j).$$

8.3 Related Work

Relatively to the atomic setting, Caragiannis et al. [52] give several results regarding the existence and the computability of efficient static tax functions with respect to pure and mixed Nash equilibria in congestion games with affine latencies. They show how to compute 2-efficient refundable taxes based on optimal strategy profiles, with respect to mixed Nash equilibria and for weighted games, and $(1+2/\sqrt{3})$-efficient universal refundable taxes with respect to pure Nash equilibria for unweighted games. These results hold for both the weighted total latency and the total latency social functions. The former is tight since it is shown that even singleton unweighted games on identical machines do not admit $(2-\varepsilon)$-efficient taxes with respect to mixed Nash equilibria (even though determining a matching lower bound for pure Nash equilibria is left as an open problem), while the latter is based on a universally static tax function and, under this restriction, it is shown to be tight either. For singleton and symmetric unweighted games, 1-efficient (i.e., optimal) refundable taxes can be computed with respect to pure Nash equilibria and this result holds even for congestion games with non-decreasing latency functions. On the negative side, several lower bounds are provided with respect to pure Nash equilibria. For singleton games with identical resources, no $(11/10-\varepsilon)$-efficient refundable taxes are possible when the players are unweighted, with the bound rising to $9/8-\varepsilon$ for weighted players and symmetric games; moreover, for non-singleton unweighted games, it becomes $6/5-\varepsilon$.

For singleton weighted games on identical resources, $(1+\sqrt{2})$-efficient non-refundable taxes with respect to mixed Nash equilibria and the weighted total cost are shown to exist. Furthermore, if the number of resources is constant, a $(1+\sqrt{2}+\varepsilon)$-efficient tax function can be computed in polynomial time. With respect to the total cost, 4-efficient non-refundable taxes with respect to mixed Nash equilibria are shown to exist, while $(6+\varepsilon)$-efficient non-refundable taxes with respect to mixed Nash equilibria can be computed in polynomial time.

Other results regarding the existence and computation of efficient taxes in congestion games under particular restrictions can be found in the works of Chen et al. [58], Fotakis and Spirakis [89], Jelinek et al. [107], Meir and Parkes [125], Ordóñez and Stier-Moses [135], Sandholm [163], Singh [168].

Taxes for non-atomic congestion games have been studied long before those defined for atomic congestion games. Indeed, Beckmann et al. [17] prove that the marginal cost taxation, introduced by Pigou [150], ensures the optimality of Nash equilibria in non-atomic congestion games, i.e. a price of anarchy equal to 1. Marginal cost taxation is defined in such a way that the latency function of each player is equal to increase she causes to the social welfare, due to her presence in the game, and can be implemented as a universal tax function; in non-atomic congestion games can be implemented as a constant (non-universal) tax function, too. In subsequent works, the existence and the computation of efficient taxes in many variants of non-atomic congestion games has been intensively studied (see, for instance, Brown and Marden [44, 45, 46], Cole et al. [70, 71], Colini-Baldeschi et al. [73], Ferguson et al. [84, 85], Fleischer et al. [87], Meir and Parkes [126], Swamy [171]).

Chapter 9
Dynamic Taxes for Congestion Games with Polynomial Latencies

In this chapter, we consider the efficiency of taxation in congestion games with polynomial latency. By resorting to a novel application of the primal-dual method, we provide interesting upper bounds with respect to a variety of different solution concepts ranging from approximate pure Nash equilibria up to approximate coarse correlated equilibria, and including also approximate one-round walks starting from the empty state. Our findings show a high beneficial effect of taxation, which increases more than linearly with the degree of the latency functions. In some cases, a tight relationship with some well-studied polynomials in Combinatorics and Number Theory, such as the Touchard and the Geometric polynomials, arises (see Section 8.2 for further details). In these cases, we can also show matching lower bounds, albeit under mild assumptions; interestingly, our upper bounds are derived by exploiting the combinatorial definition of these polynomials, while our lower bounds are constructed by relying on their analytical characterization.

Recall notations from Section 8.2. For either weighted and unweighted games, we determine $\mathcal{B}_{d+1}(1+\varepsilon)$-efficient taxes based on a social optimum with respect to ε-approximate equilibria and to both the total latency and the weighted total latency. We also show that this is the best possible efficiency achievable by a tax function based on a social optimum even in unweighted games. From a computational point of view, we show how to compute in polynomial time $(1+\xi)\mathcal{B}_{d+1}(1+\varepsilon)$-efficient taxes in unweighted games, for any arbitrary $\xi > 0$. Still for unweighted games, we determine $\mathcal{E}_{d+1}(1+\varepsilon)$-efficient taxes based on a social optimum with respect to ε-approximate one-round walks starting from the empty state and show that this is the best possible efficiency achievable by a tax function based on a social optimum. Also in this case, we show how to compute in polynomial time $(1+\xi)\mathcal{E}_{d+1}(1+\varepsilon)$-efficient taxes, for any arbitrary $\xi > 0$. Interestingly, our upper bounds are derived by exploiting the combinatorial definition of the polynomials $\mathcal{B}_i(x)$ and $\mathcal{E}_i(x)$, while our lower bounds are constructed by relying on their analytical characterization. Finally, for weighted games, we determine $\mathcal{Z}_{d+1}(1+\varepsilon)$-efficient taxes based on a social optimum with respect to ε-approximate one-round walks starting from the empty state and to the weighted total latency.

V. Bilò, C. Vinci, *Coping with Selfishness in Congestion Games*, Monographs in Theoretical Computer Science. An EATCS Series, https://doi.org/10.1007/978-3-031-30261-9_9

Furthermore, we consider taxation for non-atomic congestion games, too. In Chapter 7, we have introduced and analyzed the solution concept of ε-approximate non-atomic one-round walks, and we have shown that outcomes generated by one-round walks in polynomial congestion games are very bad in terms of social welfare, and that they are worse (even asymptotically) than the efficiency of pure Nash equilibria (i.e., the price of anarchy). By resorting to dynamic taxation, we lower the ε-competitive ratio of polynomial congestion games with maximum degree p to $(1+\varepsilon)^{p+1}(p+1)!$.

For non-refundable taxes and weighted games, we determine taxes which are $\left(\min_{x>1}\left\{(1+\varepsilon)x\mathcal{R}_d\left(\frac{(1+\varepsilon)x}{x-1}\right)\right\}\right)$-efficient and are based on a social optimum with respect to ε-approximate equilibria and to the total cost and $\mathcal{Q}_{d+1}(1+\varepsilon)$-efficient taxes based on a social optimum with respect to ε-approximate one-round walk starting from the empty state and to the weighted total cost.

For a quantitative comparison of the results presented in this chapter, see the values reported in Tables 9.1 and 9.2, where PNE and ORW stand for pure Nash equilibria and one-round walks starting from the empty state, respectively.

d	PNE without taxes		PNE with ref. taxes	PNE with non-ref. taxes
	$\mathsf{U}(\mathcal{P}(d))$, TL	$\mathsf{W}(\mathcal{P}(d))$, WTL	$\mathsf{U}(\mathcal{P}(d)) \cup \mathsf{W}(\mathcal{P}(d))$, TL or WTL	$\mathsf{W}(\mathcal{P}(d))$, TC
1	2.5	2.618	2	4
2	9.583	9.909	5	15.273
3	41.54	47.82	15	61.794
4	267.6	277	52	268.65
5	1,514	1,858	203	1,257
6	12,345	14,099	877	6,311.77
7	98,734	118,996	4,140	33,873
8	802,603	1,101,126	21,147	193,459

Table 9.1 Comparison of the efficiency of pure Nash equilibria in congestion games with polynomial latencies of maximum degree $d \leq 8$ with and without tax functions. The values for non-refundable taxes have to be compared with the unbounded price of anarchy holding for the case without taxation.

d	ORW without taxes		ORW with ref. taxes		ORW with non-ref. taxes
	$\mathsf{U}(\mathcal{P}(d))$, TL	$\mathsf{W}(\mathcal{P}(d))$, WTL	$\mathsf{U}(\mathcal{P}(d))$, WTL	$\mathsf{W}(\mathcal{P}(d))$, WTL	$\mathsf{W}(\mathcal{P}(d))$, WTC
1	4.236	7.464	3	4	6
2	37.589	90.302	13	21	57
3	527.323	1,521.451	75	148	756

Table 9.2 Comparison of the efficiency of one-round walks starting from the empty state in congestion games with polynomial latencies of maximum degree $d \leq 3$ with and without tax functions. We restrict to these values of d since no other upper bounds are known for the case without taxation.

9.1 Primal Formulation and Characteristic Inequalities

In this section, we provide the basic inequalities, to be used within the primal-dual method, characterizing the properties of ε-approximate pure Nash equilibria and ε-approximate one-round walks starting from the empty state in weighted congestion games with polynomial latencies under the application of dynamic tax functions. We assume, without loss of generality, that all players' weights are at least equal to one. This assumption can be done because of the following remark, holding even for more general classes of latency functions which are closed under abscissa-scaling:

Remark 9.1. Suppose that $\mathcal{C}$ is closed under ordinate and abscissa scaling. Then we can assume without loss of generality that all weights are at least equal to 1. Indeed, if $w^* < 1$ is the minimum weight, one can consider $\ell'_e(x) = w^*\ell(w^*x)$ in place of $\ell(x)$ as latency function of resource e, and weight $w'_i = \frac{w_i}{w^*} \geq 1$ in place of weight w_i for each $i \in N$, so that the new game is isomorphic to the initial one, and the social value of each strategy profile is invariant with respect to the considered transformation.

To establish an upper bound on the worst-case performance guarantee of a solution concept SC with respect to a social function SF in a class of games $\mathcal{G} \in \{\mathsf{W}(\mathcal{P}(d)), \mathsf{U}(\mathcal{P}(d))\}$, we use the primal-dual method. Similarly as in Chapter 3, we fix a game $\mathsf{CG} \in \mathcal{G}$, a social optimum $\boldsymbol{\sigma}^*$ and a strategy profile $\boldsymbol{\sigma} \in \mathsf{SC}(\mathsf{CG}, \delta)$, and we construct a linear program $\mathsf{LP}(\mathsf{SF}, \boldsymbol{\sigma}, \boldsymbol{\sigma}^*, \delta)$, whose optimal value is an upper bound on the efficiency of the considered tax function δ:

$$\begin{aligned} \max \quad & \mathsf{SF}(\boldsymbol{\sigma}) \\ s.t. \quad & constraints(\boldsymbol{\sigma}, \boldsymbol{\sigma}^*) \\ & \mathsf{SF}(\boldsymbol{\sigma}^*) = 1 \\ & \alpha_e \geq 0 \quad \forall e \in E, \end{aligned}$$

and the variables are the linear multipliers $\alpha_{e,h}$s of the resource latencies (as $\ell_e(x) = \sum_{h=0}^{d} \alpha_{e,h} x^h$).

Now, we determine the inequalities which are used in the following sections, in place of $constraints(\boldsymbol{\sigma}, \boldsymbol{\sigma}^*)$, to obtain upper bounds on the efficiency of dynamic taxes. We separately consider the cases of ε-approximate pure Nash equilibria and ε-approximate one-round walks.

Characteristic Inequalities for Pure Nash Equilibria

Fix a game $\mathsf{CG} \in \mathsf{W}(\mathcal{P}(d))$ and let $\delta \in \mathsf{TAX}_{\mathsf{MSP}}(\mathsf{W}(\mathcal{P}(d)))$ be a tax function for CG based on a mixed strategy profile $\mathbf{y}$, so that

$$\delta_e(w, k_e) := \sum_{h \in [d]_0} \alpha_{e,h} \delta_h(w, \mathbb{E}[k_e(\mathbf{y})], \mathbb{E}[n_e(\mathbf{y})], k_e) \tag{9.1}$$

for some functions $\delta_h : \mathbb{R}^4_{\geq 0} \to \mathbb{R}_{\geq 0}$ (as in Definition 8.1); furthermore, we assume that each $\delta_h(a,b,c,t)$ is non-decreasing in $t \geq 0$, for any fixed choice of $a,b,c \geq 0$. Let $\boldsymbol{\sigma}$ be an ε-approximate pure Nash equilibrium for (CG,δ) and let $\boldsymbol{\sigma}^*$ be a social optimum of CG. For each $i \in N$, we have

$$in(i,j) := \widehat{c}_i(\boldsymbol{\sigma}) - (1+\varepsilon)\widehat{c}_i(\boldsymbol{\sigma}_{-i}, S_{i,j}) \leq 0.$$

For the sake of brevity, we shorten $\delta_h(w, \mathbb{E}[k_e(\mathbf{y})], \mathbb{E}[n_e(\mathbf{y})], k_e)$ with $\delta_h(w,\mathbf{y},k_e)$. We have

$$\widehat{c}_i(\boldsymbol{\sigma}) = \sum_{e\in\sigma_i}\sum_{h\in[d]_0} \alpha_{e,h}\left(k_e^h + \delta_h(w_i, \mathbb{E}[k_e(\mathbf{y})], \mathbb{E}[n_e(\mathbf{y})], k_e)\right)$$

and

$$\begin{aligned}\widehat{c}_i(\boldsymbol{\sigma}_{-i},\sigma_i^*) &\leq \sum_{e\in\sigma_i^*}\sum_{h\in[d]_0} \alpha_{e,h}\left((k_e+w_i)^h + \delta_h(w_i,\mathbf{y},k_e+w_i)\right)\\ &\leq \sum_{e\in\sigma_i^*}\sum_{h\in[d]_0} \alpha_{e,h}\left((k_e+o_e)^h + \delta_h(w_i,\mathbf{y},k_e+o_e)\right),\end{aligned}$$

where the inequalities follow from the fact that $\delta_h(w_i,\mathbf{y},t)$ is non-decreasing in t.

Denote as j_i^* the index such that $\sigma_i^* = S_{i,j_i^*}$. For a tax function such that $\delta_h(w,\mathbf{y},k_e)$ can be written as $w\delta_h(\mathbf{y},k_e)$ (i.e., it is linearly dependent on the weight w), by computing the inequality $\sum_{i\in N} in(i,j_i^*)$, we obtain

$$\begin{aligned}&\sum_{e\in E}\sum_{h\in[d]_0} \alpha_{e,h}\left(n_e k_e^h + k_e\delta_h(\mathbf{y},k_e)\right)\\ &\qquad - (1+\varepsilon)\sum_{e\in E}\sum_{h\in[d]_0} \alpha_{e,h}\left(n_e^*(k_e+o_e)^h + o_e\delta_h(\mathbf{y},k_e+o_e)\right)\\ &\leq \sum_{i\in N} in(i,j_i^*) \leq 0. \end{aligned} \tag{9.2}$$

For a tax function such that $\delta_h(w,\mathbf{y},k_e)$ can be written as $\delta_h(\mathbf{y},k_e)$ (i.e., it does not depend on the weight w), by computing the inequality $\sum_{i\in N} w_i in(i,j_i^*)$, we obtain

$$\begin{aligned}&\sum_{e\in E}\sum_{h\in[d]_0} \alpha_{e,h}\left(k_e^{h+1} + k_e\delta_h(\mathbf{y},k_e)\right)\\ &\qquad - (1+\varepsilon)\sum_{e\in E}\sum_{h\in[d]_0} \alpha_{e,h}\left(o_e(k_e+o_e)^h + o_e\delta_h(\mathbf{y},k_e+o_e)\right)\\ &\leq \sum_{i\in N} w_i in(i,j_i^*) \leq 0. \end{aligned} \tag{9.3}$$

Finally, for an unweighted game $\mathsf{CG} \in \mathsf{U}(\mathcal{P}(d))$ and a tax function in $\mathsf{TAX}_{\mathsf{MSP}}(\mathsf{U}(\mathcal{P}(d)))$ based on a mixed profile $\mathbf{y}$, we have

$$\widehat{c}_i(\boldsymbol{\sigma}) = \sum_{e\in\sigma_i}\sum_{h\in[d]_0} \alpha_{e,h}\left(k_e^h + \delta_h(\mathbb{E}[k_e(\mathbf{y})], k_e)\right)$$

and

$$\widehat{c}_i(\boldsymbol{\sigma}_{-i}, S_{i,j}) \leq \sum_{e \in S_{i,j}} \sum_{h \in [d]_0} \alpha_{e,h} \left((k_e+1)^h + \delta_h(\mathbb{E}[k_e(\mathbf{y})], k_e+1) \right)$$

for any $j \in [m_i]$, so that, by computing the inequality $\sum_{i \in N} \sum_{j \in [m_i]} y_{i,j} in(i,j)$, we get

$$\begin{aligned} &\sum_{e \in E} \sum_{h \in [d]_0} \alpha_{e,h} \left(k_e^{h+1} + k_e \delta_h(\mathbb{E}[k_e(\mathbf{y})], k_e) \right) \\ &\quad - (1+\varepsilon) \sum_{e \in E} \sum_{h \in [d]_0} \alpha_{e,h} \mathbb{E}[k_e(\mathbf{y})] \left((k_e+1)^h + \delta_h(\mathbb{E}[k_e(\mathbf{y})], k_e+1) \right) \\ &\leq \sum_{i \in N} \sum_{j \in [m_i]} y_{i,j} in(i,j) \leq 0. \end{aligned} \tag{9.4}$$

Characteristic Inequalities for One-Round Walks

Consider a game $\mathsf{CG} \in \mathsf{W}(\mathcal{P}(d))$ and let $\delta \in \mathsf{TAX}_{\mathsf{MSP}}(\mathsf{W}(\mathcal{P}(d)))$ be a tax function for CG based on a mixed strategy profile $\mathbf{y}$ and defined as in (9.1). Let $\boldsymbol{\sigma}$ be the strategy profile generated by an ε-approximate one-round walk starting from the empty state $\boldsymbol{\tau} = (\boldsymbol{\sigma}^0, \ldots, \boldsymbol{\sigma}^n)$ for game (CG, δ). By the definition of $\boldsymbol{\tau}$, for each $i \in N$ and $j \in [m_i]$, we have

$$\widetilde{in}(i,j) := \widehat{c}_i(\boldsymbol{\sigma}_{-i}^{i-1}, \sigma_i) - (1+\varepsilon)\widehat{c}_i(\boldsymbol{\sigma}_{-i}^{i-1}, S_{i,j}) \leq 0.$$

We have

$$\widehat{c}_i(\boldsymbol{\sigma}^i) = \sum_{e \in \sigma_i} \sum_{h \in [d]_0} \alpha_{e,h} \left((k_e(\boldsymbol{\sigma}^i))^h + \delta_h(w_i, \mathbf{y}, k_e(\boldsymbol{\sigma}^i)) \right)$$

and

$$\begin{aligned} \widehat{c}_i(\boldsymbol{\sigma}_{-i}^{i-1}, S_{i,j_i^*}) &= \sum_{e \in \sigma_i^*} \sum_{h \in [d]_0} \alpha_{e,h} \left((k_e(\boldsymbol{\sigma}^{i-1}) + w_i)^h + \delta_h(w_i, \mathbf{y}, k_e(\boldsymbol{\sigma}^{i-1}) + w_i) \right) \\ &\leq \sum_{e \in \sigma_i^*} \sum_{h \in [d]_0} \alpha_{e,h} \left((k_e + o_e)^h + \delta_h(w_i, \mathbf{y}, k_e + o_e) \right), \end{aligned}$$

where the inequality follows from the fact that $\delta_h(w_i, \mathbf{y}, t)$ is non-decreasing in t.

For a tax function $\delta_h(w, \mathbf{y}, k_e)$ that can be written as $\delta_h(\mathbf{y}, k_e)$, by computing the inequality $\sum_{i \in N} w_i \widetilde{in}(i, j_i^*)$, and by resorting to similar arguments as in Theorem 4.1 of Chapter 4, we obtain

$$\begin{aligned} 0 &\geq \sum_{i \in N} w_i \widetilde{in}(i, j_i^*) \\ &\geq \sum_{e \in E} \sum_{h \in [d]_0} \alpha_{e,h} \sum_{i \in N} w_i \left(k_e(\boldsymbol{\sigma}^i) + \delta_h(\mathbf{y}, k_e(\boldsymbol{\sigma}^i)) \right) \chi_{\sigma_i}(e) \end{aligned}$$

$$
\begin{aligned}
&- (1+\varepsilon) \sum_{e\in E} \sum_{h\in[d]_0} \alpha_{e,h} \left(o_e(k_e+o_e)^h + o_e \delta_h(\mathbf{y}, k_e+o_e) \right) \\
&\geq \sum_{e\in E} \sum_{h\in[d]_0} \alpha_{e,h} \int_0^{k_e} (t^h + \delta_h(\mathbf{y},t))dt \\
&\quad - (1+\varepsilon) \sum_{e\in E} \sum_{h\in[d]_0} \alpha_{e,h} \left(o_e(k_e+o_e)^h + o_e \delta_h(\mathbf{y}, k_e+o_e) \right). \qquad (9.5)
\end{aligned}
$$

Finally, for a tax function in $\mathsf{TAX}_{\mathsf{MSP}}(\mathsf{U}(\mathcal{P}(d)))$ based on a mixed profile $\mathbf{y}$, we have

$$
\widehat{c}_i(\boldsymbol{\sigma}^i) = \sum_{e\in\sigma_i} \sum_{h\in[d]_0} \alpha_{e,h} \left((k_e(\boldsymbol{\sigma}^i))^h + \delta_h(k_e(\mathbb{E}[k_e(\mathbf{y})], k_e(\boldsymbol{\sigma}^i)) \right)
$$

and

$$
\widehat{c}_i(\boldsymbol{\sigma}^{i-1}_{-i}, S_{i,j}) \leq \sum_{e\in\sigma_i^*} \sum_{h\in[d]_0} \alpha_{e,h} \left((k_e(\boldsymbol{\sigma}^{i-1})+1)^h + \delta_h(\mathbb{E}[k_e(\mathbf{y})], k_e(\boldsymbol{\sigma}^{i-1})+1) \right)
$$

so that, by computing the inequality $\sum_{i\in N}\sum_{j\in[m_i]} y_{i,j}\widetilde{in}(i,j)$, we get

$$
\begin{aligned}
0 &\geq \sum_{i\in N} \sum_{j\in[m_i]} y_{i,j}\widetilde{in}(i,j) \\
&\geq \sum_{e\in E} \sum_{h\in[d]_0} \alpha_{e,h} \sum_{i\in N} \left((k_e(\boldsymbol{\sigma}^i))^h + \delta_h(\mathbb{E}[k_e(\mathbf{y})], k_e(\boldsymbol{\sigma}^i)) \right) \chi_{\sigma_i}(e) \\
&\quad - (1+\varepsilon) \sum_{e\in E} \sum_{h\in[d]_0} \alpha_{e,h} \mathbb{E}[k_e(\mathbf{y})] \left((k_e+1)^h + \delta_h(\mathbb{E}[k_e(\mathbf{y})], k_e+1) \right) \\
&= \sum_{e\in E} \sum_{h\in[d]_0} \alpha_{e,h} \sum_{l=1}^{n_e} (l^h + \delta_h(\mathbb{E}[k_e(\mathbf{y})], l)) \\
&\quad - (1+\varepsilon) \sum_{e\in E} \sum_{h\in[d]_0} \alpha_{e,h} \mathbb{E}[k_e(\mathbf{y})] \left((k_e+1)^h + \delta_h(\mathbb{E}[k_e(\mathbf{y})], k_e+1) \right). \qquad (9.6)
\end{aligned}
$$

9.2 Refundable Taxes for Atomic Games

9.2.1 Taxes for Nash Equilibria

As a first result, we show how to define an efficient dynamic tax function based on a social optimum with respect to ε-approximate pure Nash equilibria for weighted games.

Remark 9.2. In Remark 4.1, we have pointed out that, as consequence of the results of Roughgarden [158] and Bilò [42], the efficiency of pure Nash equilibria is as high as that of its generalizations (from mixed to coarse-correlated equilibria). The same fact holds even under the type of taxes we adopt, despite we will simply refer to the restricted case of pure Nash equilibria.

Theorem 9.1. *Fix an integer* $d \geq 1$. *For any* $\mathsf{CG} \in \mathsf{W}(\mathcal{P}(d))$, *there exists* $\delta \in \mathsf{TAX}_{\mathsf{OPT}}(\mathsf{W}(\mathcal{P}(d)))$ *such that* δ *is a* $\mathcal{B}_{d+1}(1+\varepsilon)$*-efficient refundable tax with respect to* ε*-approximate pure Nash equilibria and to both the total latency and the weighted total latency.*

Proof. For an integer $d \geq 1$, fix a weighted congestion game $\mathsf{CG} \in \mathsf{W}(\mathcal{P}(d))$ and let $\boldsymbol{\sigma}^*$ be a social optimum for CG with respect to the social function SF. Throughout this proof, we set $q_e^* = n_e^*$ for $\mathsf{SF} = \mathsf{TL}$ and $q_e^* = o_e$ for $\mathsf{SF} = \mathsf{WTL}$. Similarly, for a given strategy profile $\boldsymbol{\sigma}$, we set $q_e = n_e$ for $\mathsf{SF} = \mathsf{TL}$ and $q_e = k_e$ for $\mathsf{SF} = \mathsf{WTL}$. Consider a dynamic tax function $\delta \in \mathsf{TAX}_{\mathsf{OPT}}(\mathsf{W}(\mathcal{P}(d)))$ based on $\boldsymbol{\sigma}^*$ such that $\delta_h(w, o_e, n_e^*, k_e) := w\delta_h(o_e, n_e^*, k_e)$ for $\mathsf{SF} = \mathsf{TL}$ and $\delta_h(w, o_e, n_e^*, k_e) := \delta_h(o_e, n_e^*, k_e)$ for $\mathsf{SF} = \mathsf{WTL}$, where

$$\delta_h(o_e, n_e^*, k_e) := \sum_{i=1}^{h} \delta_{h,i}(o_e, n_e^*, k_e)$$

with

$$\delta_{h,i}(o_e, n_e^*, k_e) := \sum_{j=1}^{i} \mathcal{S}(i,j)(1+\varepsilon)^j q_e^* o_e^{i-1}(k_e + j o_e)^{h-i}.$$

Note that $\delta_h(o_e, n_e^*, k_e)$ is a polynomial function in k_e with non-negative coefficients, therefore it is non-negative and non-decreasing in k_e so that we can use the inequalities developed in Section 9.1. Now let $\boldsymbol{\sigma}$ be an ε-approximate pure Nash equilibrium for (CG, δ). By applying the primal-dual method defined in Chapter 3, we get the linear program $\mathsf{LP} := \mathsf{LP}(\mathsf{SF}, \boldsymbol{\sigma}, \boldsymbol{\sigma}^*, \delta)$ described in Section 9.1, with $constraints(\boldsymbol{\sigma}, \boldsymbol{\sigma}^*)$ equal to inequality (9.2) for $\mathsf{SF} = \mathsf{TL}$ and to inequality (9.3) for $\mathsf{SF} = \mathsf{WTL}$.

The related dual program DLP is the following one (we associate the dual variables x and γ with the first and second constraint of LP, respectively):

$$\begin{aligned} \min \quad & \gamma \\ s.t. \quad & q_e k_e^h \leq \gamma q_e^* o_e^h + x\left(q_e k_e^h + k_e \delta_h(o_e, n_e^*, k_e)\right) \\ & \quad - x(1+\varepsilon)\left(q_e^*(k_e + o_e)^h + o_e \delta_h(o_e, n_e^*, k_e + o_e)\right) \quad \forall e \in E, h \in [d]_0 \\ & x \geq 0. \end{aligned}$$

Now we show that, for each $e \in E$ and $h \in [d]_0$, the dual constraint obtained by setting $x = 1$ and $\gamma = \mathcal{B}_{d+1}(1+\varepsilon)$ is always satisfied by the proposed tax function δ. To this aim, we exploit the following result.

Lemma 9.1. *For each* $e \in E$ *and* $h \in [d]_0$, *we have*

$$k_e \delta_h(o_e, n_e^*, k_e) - (1+\varepsilon)\left(q_e^*(k_e + o_e)^h + o_e \delta_h(o_e, n_e^*, k_e + o_e)\right)$$

$$= -\sum_{j=1}^{h+1} \mathcal{S}(h+1,j)(1+\varepsilon)^j q_e^* o_e^h. \quad (9.7)$$

Proof (of the lemma). To show the claim of Lemma 9.1, one can observe that the considered tax function δ_h has been defined in such a way that the left-hand part of (9.7) is constant with respect to k_e. In particular, define $T_{h,i}(o_e, n_e^*, k_e) := \sum_{r=i}^{h} \delta_{h,r}(o_e, n_e^*, k_e)$ so that

$$T_{h,i}(o_e,n_e^*,k_e) = \delta_{h,i}(o_e,n_e^*,k_e) + T_{h,i+1}(o_e,n_e^*,k_e) \quad (9.8)$$

for each $i \in [h+1]$. One can prove by induction on $i \in [h+1]$ that

$$k_e\delta_h(o_e,n_e^*,k_e) - (1+\varepsilon)\left(q_e^*(k_e+o_e)^h + o_e\delta_h(o_e,n_e^*,k_e+o_e)\right)$$
$$= k_e T_{h,i}(o_e,n_e^*,k_e) - \delta_{h+1,i}(o_e,n_e^*,k_e) - (1+\varepsilon)o_e T_{h,i}(o_e,n_e^*,k_e+o_e). \quad (9.9)$$

By using $i = h+1$ in (9.9), we get (observe that $T_{h,h+1} = 0$)

$$k_e\delta_h(o_e,n_e^*,k_e) - (1+\varepsilon)\left(q_e^*(k_e+o_e)^h + o_e\delta_h(o_e,n_e^*,k_e+o_e)\right)$$
$$= -\delta_{h+1,h+1}(o_e,n_e^*,k_e)$$
$$= -\sum_{j=1}^{h+1} \mathcal{S}(h+1,j)(1+\varepsilon)^j q_e^* o_e^h,$$

and this shows Lemma 9.1. □

Finally, the claim of the theorem follows since $\sum_{j=1}^{h+1} \mathcal{S}(h+1,j)(1+\varepsilon)^j q_e^* o_e^h = q_e^* o_e^h \mathcal{B}_{h+1}(1+\varepsilon)$ and $\mathcal{B}_{h+1}(1+\varepsilon) \leq \mathcal{B}_{d+1}(1+\varepsilon)$ for each $h \in [d]_0$, that is, the dual constraint of DLP is satisfied. □

The upper bound given in Theorem 9.1 is only an existential result, since we do not know how to compute in polynomial time the social optimum of a given weighted congestion game with polynomial latencies. We now show that, if we restrict to unweighted games, efficient taxes asymptotically matching the lower bound given in Theorem 9.3 can be computed in polynomial time by solving a convex optimization problem.

Theorem 9.2. *Fix an integer $d \geq 1$ and an arbitrarily small value $\xi > 0$. For any $\mathsf{CG} \in \mathsf{U}(\mathcal{P}(d))$, it is possible to compute in polynomial time a tax function $\delta \in \mathsf{TAX}_{\mathsf{MSP}}(\mathsf{U}(\mathcal{P}(d)))$ such that δ is a $(1+\xi)\mathcal{B}_{d+1}(1+\varepsilon)$-efficient refundable tax with respect to ε-approximate pure Nash equilibria and to the total latency.*

Proof. Fix an integer $d \geq 1$, an arbitrarily small value $\xi > 0$ and an unweighted congestion game $\mathsf{CG} \in \mathsf{U}(\mathcal{P}(d))$.

Consider a dynamic tax function $\delta \in \mathsf{TAX}_{\mathsf{MSP}}(\mathsf{U}(\mathcal{P}(d)))$ based on a mixed strategy profile $\mathbf{y} \in \boldsymbol{\Delta}$ such that

$$\delta_h(\mathbb{E}[k_e(\mathbf{y})], k_e) := \sum_{i=1}^{h} \delta_{h,i}(\mathbb{E}[k_e(\mathbf{y})], k_e)$$

with

$$\delta_{h,i}(\mathbb{E}[k_e(\mathbf{y})], k_e) := \sum_{j=1}^{i} \mathcal{S}(i,j)\left((1+\varepsilon)\mathbb{E}[k_e(\mathbf{y})]\right)^j (k_e+j)^{h-i}.$$

By exploiting the same arguments as in the proof of Lemma 9.1, we get

$$k_e\delta_h(\mathbb{E}[k_e(\mathbf{y})], k_e) - (1+\varepsilon)\mathbb{E}[k_e(\mathbf{y})]\left((k+1)^h + \delta_h(\mathbb{E}[k_e(\mathbf{y})], k_e+1)\right) = -\sum_{j=1}^{h+1} \mathcal{S}(h+1,j)(1+\varepsilon)^j \mathbb{E}[k_e(\mathbf{y})]^j$$

so that

$$\begin{aligned}&\sum_{e\in E}\sum_{h\in[d]_0} \alpha_{e,h} k_e \delta_h(\mathbb{E}[k_e(\mathbf{y})], k_e) \\ &\quad - \sum_{e\in E}\sum_{h\in[d]_0} \alpha_{e,h}(1+\varepsilon)\mathbb{E}[k_e(\mathbf{y})]\left((k+1)^h + \delta_h(\mathbb{E}[k_e(\mathbf{y})], k_e+1)\right) \\ &= -\sum_{e\in E}\sum_{h\in[d]_0} \alpha_{e,h}\mathcal{B}_{h+1}((1+\varepsilon)\mathbb{E}[k_e(\mathbf{y})]).\end{aligned} \tag{9.10}$$

Equality (9.10) combined with inequality (9.4) gives

$$\sum_{e\in E}\sum_{h\in[d]_0} \alpha_{e,h}\left(k_e^{h+1} - \mathcal{B}_{h+1}((1+\varepsilon)\mathbb{E}[k_e(\mathbf{y})])\right) \le 0.$$

Consider now the following convex optimization problem:

$$\begin{aligned}\min\ & \sum_{e\in E}\sum_{h\in[d]_0} \alpha_{e,h}\mathcal{B}_{h+1}\left((1+\varepsilon)\mathbb{E}[k_e(\mathbf{y})]\right) \\ s.t.\ & \sum_{j=1}^{m_i} y_{i,j} = 1 && \forall i \in N \\ & \mathbb{E}[k_e(\mathbf{y})] = \sum_{i\in N}\sum_{j\in[m_i]:e\in S_{i,j}} y_{i,j} && \forall e \in E \\ & y_{i,j} \ge 0 && \forall i \in N, j \in [m_i]\end{aligned}$$

and let $\overline{\mathbf{y}} = (\overline{\mathbf{y}}_i, \ldots, \overline{\mathbf{y}}_n)$ be a $(1+\xi)$-approximation of its optimal solution. Such a solution can be computed in polynomial time, by exploiting an interior-point method for convex optimization (see [43]). Furthermore, if $d = 1$, we have a stronger result: one can resort to convex quadratic programming (see [69]) to find an optimal solution in polynomial time (i.e. for $\xi = 0$). By the property of $\overline{\mathbf{y}}$, it follows that

$$\sum_{e\in E}\sum_{h\in[d]_0} \alpha_{e,h}\mathcal{B}_{h+1}((1+\varepsilon)\mathbb{E}[k_e(\overline{\mathbf{y}})]) \le \sum_{e\in E}\sum_{h\in[d]_0} \alpha_{e,h}(1+\xi)\mathcal{B}_{h+1}((1+\varepsilon)o_e).$$

Hence, in order to evaluate the efficiency of the tax function induced by $\overline{\mathbf{y}}$ within the primal-dual method, we can exploit the inequality

$$\sum_{e\in E}\sum_{h\in[d]_0} \alpha_{e,h}\left(k_e^{h+1} - (1+\xi)\mathcal{B}_{h+1}((1+\varepsilon)o_e)\right) \leq 0$$

in place of

$$\sum_{e\in E}\sum_{h\in[d]_0} \alpha_{e,h}\left(k_e^{h+1} - \mathcal{B}_{h+1}((1+\varepsilon)\mathbb{E}[k_e(\overline{\mathbf{y}})])\right) \leq 0.$$

By applying the primal-dual method, we get the following linear program LP:

$$\begin{aligned}
\max\ & \mathsf{WTL}(\boldsymbol{\sigma}) \\
s.t.\ & \sum_{e\in E}\sum_{h\in[d]_0} \alpha_{e,h}\left(k_e^{h+1} - (1+\xi)\mathcal{B}_{h+1}((1+\varepsilon)o_e)\right) \leq 0 \\
& \mathsf{WTL}(\boldsymbol{\sigma}^*) = 1 \\
& \alpha_{e,h} \geq 0 \qquad\qquad \forall e\in E, h\in[d]_0.
\end{aligned}$$

The related dual program is the following one (we associate the dual variables x and γ with the first and second constraint of LP, respectively):

$$\begin{aligned}
\min\ & \gamma \\
s.t.\ & k_e^{h+1} \leq \gamma o_e^{h+1} + x\left(k_e^{h+1} - (1+\xi)\mathcal{B}_{h+1}((1+\varepsilon)o_e)\right) \ \forall e\in E, h\in[d]_0 \\
& x \geq 0.
\end{aligned}$$

Now since $\mathcal{B}_{h+1}((1+\varepsilon)o_e) \leq o_e^{h+1}\mathcal{B}_{h+1}(1+\varepsilon)$, for each $e\in E$ and $h\in[d]_0$, the dual constraint obtained by setting $x=1$ and $\gamma=(1+\xi)\mathcal{B}_{d+1}(1+\varepsilon)$ is always satisfied by the proposed tax function δ. Again the claim follows since $\mathcal{B}_{h+1}(1+\varepsilon) \leq \mathcal{B}_{d+1}(1+\varepsilon)$ for each $h\in[d]_0$. □

Now, we exhibit a matching lower bound holding for dynamic tax functions based on a social optimum.

Theorem 9.3. *Fix an integer $d\geq 1$. For any fixed $\xi>0$, there exists $\mathsf{CG}\in\mathsf{U}(\mathcal{P}(d))$ such that, for any $\delta\in\mathsf{TAX}_{\mathsf{OPT}}(\mathsf{U}(\mathcal{P}(d)))$, δ is not a $(\mathcal{B}_{d+1}(1+\varepsilon)-\xi)$-efficient refundable tax for* CG *with respect to ε-approximate pure Nash equilibria and to the total latency.*

Proof. Fix an integer $d\geq 1$, a dynamic tax functions $\delta\in\mathsf{TAX}_{\mathsf{OPT}}(\mathsf{U}(\mathcal{P}(d)))$ and a value $\xi>0$. Game CG has $n+1$ players and a set of resources E such that $E=\{E_0,E_1,\ldots,E_n\}$ with $E_j=\{e_{j,1},\ldots,e_{j,n}\}$ for each $j\in[n]_0$, that is, E is partitioned into $n+1$ groups each of which contains n resources. All the resources in the same group E_j have the same latency function $\ell_j(x)=\alpha_j x^d$ where $\alpha_j=\frac{(1+\varepsilon)^j}{j!}$ for $j\in[n-1]_0$ and $\alpha_n=\frac{(1+\varepsilon)^n}{(n-1)(n-1)!}$. For each $i\in[n]$, $\Sigma_i=\{S_{i,1},S_{i,2}\}$ with $S_{i,1}:=\bigcup_{j=1}^{n-1}\bigcup_{q=1}^{j}\{e_{j,i+q}\}\cup\bigcup_{q=1}^{n-1}\{e_{n,i+q}\}$ and $S_{i,2}:=\bigcup_{j=0}^{n-1}\{e_{j,i}\}$, where the resource indexes are assumed to be cyclical, i.e., $e_{j,n+1}=e_{j,1}$, while $\Sigma_{n+1}=\{S_{n+1,1}\}$ with $S_{n+1,1}=E_n$. The lower bound instance for $n=6$ is depicted in Figure 9.1.

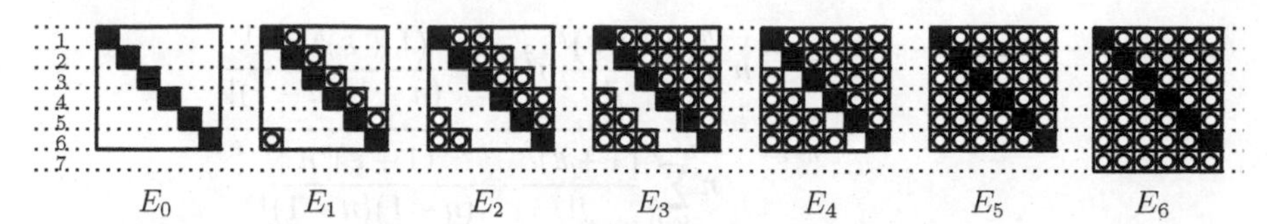

Fig. 9.1 The lower bound instance of Theorem 9.3 for $n = 6$. On the abscissa we have the resources (organized in groups $E_0, E_1, \ldots, E_n$), and on the ordinate we have the players. For any player $i \in [7]_1$, the squares containing circles and having position i on the ordinate represent the resources belonging to strategy $S_{i,1}$, and the set of black squares having position i represent the resources belonging to $S_{i,2}$.

First, we show that the unique social optimum with respect to the total latency is given by the strategy profile $\boldsymbol{\sigma}^* := (S_{1,2}, S_{2,2}, \ldots, S_{n,2}, S_{n+1,1})$. To this aim, observe that $c_i(\boldsymbol{\sigma}^*) = \sum_{j=0}^{n-1} \alpha_j$ for each $i \in [n]$ which implies $\mathsf{TL}(\boldsymbol{\sigma}^*) = n\sum_{j=0}^{n-1} \alpha_j + n\alpha_n$. Consider now any other strategy profile $\boldsymbol{\sigma}' \neq \boldsymbol{\sigma}^*$ and note that in $\boldsymbol{\sigma}'$ there always exists at least a resource e such that $k_e > 1$. This implies that $\mathsf{TL}(\boldsymbol{\sigma}') > \sum_{i\in[n+1]} \sum_{e\in\sigma'_i} \ell_e(1)$. By $\sum_{e\in\sigma'_i} \ell_e(1) = \min\{\sum_{j=0}^{n-1} \alpha_j, \sum_{j=1}^{n-1} j\alpha_j + (n-1)\alpha_n\} = \sum_{j=0}^{n-1} \alpha_j$, it follows that $\mathsf{TL}(\boldsymbol{\sigma}^*) < \mathsf{TL}(\boldsymbol{\sigma}')$.

Consider now the strategy profile $\boldsymbol{\sigma} = (S_{1,1}, S_{2,1}, \ldots, S_{n,1}, S_{n+1,1})$. For each $i \in [n]$, we have

$$\begin{aligned}
\widehat{c}_i(\boldsymbol{\sigma}) &= \sum_{j=1}^{n-1} j\alpha_j(j^d + \delta_d(1,j)) + (n-1)\alpha_n(n^d + \delta_d(1,n)) \\
&= (1+\varepsilon)\sum_{j=1}^{n} \alpha_{j-1}(j^d + \delta_d(1,j)) \\
&= (1+\varepsilon)\sum_{j=0}^{n-1} \alpha_j((j+1)^d + \delta_d(1,j+1)),
\end{aligned}$$

whereas

$$\widehat{c}_i(\boldsymbol{\sigma}_{-i}, S_{i,2}) = \sum_{j=0}^{n-1} \alpha_j((j+1)^d + \delta_d(1,j+1)).$$

Hence, it follows that $\boldsymbol{\sigma}$ is an ε-approximate pure Nash equilibrium for CG. Let us now evaluate the ratio $\mathsf{TL}(\boldsymbol{\sigma})/\mathsf{TL}(\boldsymbol{\sigma}^*)$ to obtain a lower bound on the efficiency of δ. By the fact that $c_i(\boldsymbol{\sigma}) = \sum_{j=1}^{n} j\alpha_j j^d$ and $c_i(\boldsymbol{\sigma}^*) = \sum_{j=0}^{n-1} \alpha_j$ for each $i \in [n]$, $c_{n+1}(\boldsymbol{\sigma}) = n^{d+1}\alpha_n$ and $c_n(\boldsymbol{\sigma}^*) = n\alpha_n$, we get

$$\lim_{n\to\infty} \frac{\mathsf{TL}(\boldsymbol{\sigma})}{\mathsf{TL}(\boldsymbol{\sigma}^*)} = \lim_{n\to\infty} \frac{\sum_{i=1}^{n}\sum_{j=1}^{n} j\alpha_j j^d + n^{d+1}\alpha_n}{\sum_{i=1}^{n}\sum_{j=0}^{n-1} \alpha_j + n\alpha_n}$$

$$= \lim_{n\to\infty} \frac{n\sum_{j=1}^{n} \frac{(1+\varepsilon)^j}{j!} j^{d+1} + \frac{(1+\varepsilon)^n n^{d+1}}{(n-1)(n-1)!}}{n\sum_{j=0}^{n-1} \frac{(1+\varepsilon)^j}{j!} + \frac{(1+\varepsilon)^n n}{(n-1)(n-1)!}}$$

$$= \lim_{n\to\infty} \frac{\sum_{j=0}^{n} \frac{(1+\varepsilon)^j}{j!} j^{d+1}}{e^{1+\varepsilon}}.$$

The claim follows by applying Dobiński's formula (8.3) to the last quantity. □

9.2.2 Taxes for One-Round Walks

We now move our attention to the determination of efficient tax functions with respect to ε-approximate one-round walks starting from the empty state.

Theorem 9.4. *Fix an integer $d \geq 1$. For any $\mathsf{CG} \in \mathsf{U}(\mathcal{P}(d))$, there exists $\delta \in \mathsf{TAX}_{\mathsf{OPT}}(\mathsf{U}(\mathcal{P}(d)))$ such that δ is a $\mathcal{E}_{d+1}(1+\varepsilon)$-efficient refundable tax with respect to ε-approximate one-round walks starting from the empty state and to the total latency. Moreover, for any $\xi > 0$, it is possible to compute in polynomial time a tax function $\delta \in \mathsf{TAX}_{\mathsf{MSP}}(\mathsf{U}(\mathcal{P}(d)))$ such that δ is a $(1+\xi)\mathcal{E}_{d+1}(1+\varepsilon)$-efficient refundable tax with respect to ε-approximate one-round walks starting from the empty state and to the total latency.*

Proof. For an integer $d \geq 1$, fix an unweighted congestion game $\mathsf{CG} \in \mathsf{U}(\mathcal{P}(d))$ and let $\boldsymbol{\sigma}^*$ be a social optimum for CG with respect to the social function WTL. Consider a dynamic tax function $\delta \in \mathsf{TAX}_{\mathsf{OPT}}(\mathsf{U}(\mathcal{P}(d)))$ based on $\boldsymbol{\sigma}^*$ such that

$$\delta_h(o_e,k_e) := \sum_{i=0}^{h} ((1+\varepsilon)o_e)^i \Delta^{i+1}\left[(k_e-1)^{h+1}\right] - k_e^h.$$

First of all, we show that $\delta_h(o_e,k_e) \geq 0$ for each pair of integers (o_e,k_e) such that $o_e \geq 0$ and $k_e \geq 1$ and that $\delta_h(o_e,k_e)$ is non-decreasing in $k_e \geq 1$, so that δ is indeed a tax function and the inequalities derived in Section 9.1 can be applied. We get

$$\begin{aligned}
\delta_h(o_e,k_e) &= \sum_{i=0}^{h} ((1+\varepsilon)o_e)^i \Delta^{i+1}\left[(k_e-1)^{h+1}\right] - k_e^h \\
&= \Delta^1\left[(k_e-1)^{h+1}\right] - k_e^h + \sum_{i=1}^{h} ((1+\varepsilon)o_e)^i \Delta^{i+1}\left[(k_e-1)^{h+1}\right] \\
&= k_e^{h+1} - (k_e-1)^{h+1} - k_e^h + \sum_{i=1}^{h} ((1+\varepsilon)o_e)^i \Delta^{i+1}\left[(k_e-1)^{h+1}\right]
\end{aligned}$$

$$= (k_e^h - (k_e - 1)^h)(k_e - 1) + \sum_{i=1}^{h} ((1+\varepsilon)o_e)^i \Delta^{i+1} \left[(k_e - 1)^{h+1}\right].$$

Since $(k_e^h - (k_e - 1)^h)(k_e - 1)$ and $\sum_{i=1}^h ((1+\varepsilon)o_e)^i \Delta^{i+1} \left[(k_e - 1)^{h+1}\right]$ are non-negative and non-decreasing in $k_e \geq 1$, the claim holds.

Now let $\boldsymbol{\sigma}$ be the strategy profile generated by an ε-approximate one-round walk starting from the empty state for (CG, δ). By applying the primal-dual method, we get the linear program $\mathsf{LP} := \mathsf{LP}(\mathsf{TL}, \boldsymbol{\sigma}, \boldsymbol{\sigma}^*, \delta)$ described in Section 9.1, with $constraints(\boldsymbol{\sigma}, \boldsymbol{\sigma}^*)$ equal to inequality (9.6) where we set $\mathbf{y} = \boldsymbol{\sigma}^*$ so that $\mathbb{E}[k_e(\mathbf{y})] = o_e$.

The related dual program DLP is the following one (we associate the dual variables x and γ with the first and second constraint of LP, respectively):

$$\begin{aligned}
&\min \gamma \\
&s.t.\ \ k_e^{h+1} \leq \gamma o_e^{h+1} + x\left(\sum_{j=1}^{k_e} j^h + \delta_h(o_e, j)\right) \\
&\qquad -x(1+\varepsilon)o_e\left((k_e+1)^h + \delta_h(o_e, k_e+1)\right) \ \forall e \in E, h \in [d]_0 \\
&\qquad x \geq 0
\end{aligned}$$

Now we show that, for each $e \in E$ and $h \in [d]_0$, the dual constraint obtained by setting $x = 1$ and $\gamma = \mathcal{E}_{d+1}(1+\varepsilon)$ is always satisfied by the proposed tax function δ. To this aim, we exploit the following result, that is similar to Lemma 9.1.

Lemma 9.2. *For each $e \in E$ and $h \in [d]_0$, $-k_e^{h+1} + \sum_{j=1}^{k_e}(j^h + \delta_h(o_e, j)) - (1+\varepsilon)o_e\left((k_e+1)^h + \delta_h(o_e, k_e+1)\right) = -\sum_{j=1}^{h+1} \mathcal{S}(h+1, j) j!((1+\varepsilon)o_e)^j$.*

Proof (of the lemma). Define

$$T_{h,i}(o_e, k_e) := \sum_{r=i}^{h} ((1+\varepsilon)o_e)^r \Delta^{r+1} \left[(k_e - 1)^{h+1}\right]$$

so that, for each $i \in [h+1]_0$,

$$T_{h,0}(o_e, k_e) = \delta_h(o_e, k_e) + k_e^h, \tag{9.11}$$

$$T_{h,i}(o_e, k_e) = ((1+\varepsilon)o_e)^i \Delta^{i+1} \left[(k_e - 1)^{h+1}\right] + T_{h,i+1}(o_e, k_e), \tag{9.12}$$

where Δ^i has been defined in Section 8.2. One can prove by induction on $i \in [h+1]_0$ that

$$-k_e^{h+1} + \sum_{j=1}^{k_e} (j^h + \delta_h(o_e, j)) - (1+\varepsilon)o_e\left((k_e+1)^h + \delta_h(o_e, k_e+1)\right)$$

$$= -\sum_{j=1}^{i-1} ((1+\varepsilon)o_e)^j \Delta^j \left[x^{h+1}\right]\Big|_{x=0} - ((1+\varepsilon)o_e)^i \Delta^i \left[k_e^{h+1}\right]$$

$$+\sum_{j=1}^{k_e} T_{h,i}(o_e, j) - (1+\varepsilon)o_e T_{h,i}(o_e, k_e+1). \quad (9.13)$$

By using $i = h+1$ in (9.13), we get

$$\begin{aligned}
&-\sum_{j=1}^{h}((1+\varepsilon)o_e)^j \Delta^j \left[x^{\underline{h+1}}\right]|_{x=0} - ((1+\varepsilon)o_e)^{h+1}\Delta^{h+1}\left[k_e^{\underline{h+1}}\right] \\
&= -\sum_{j=1}^{h}((1+\varepsilon)o_e)^j \Delta^j \left[x^{\underline{h+1}}\right]|_{x=0} - ((1+\varepsilon)o_e)^{h+1}\Delta^{h+1}\left[x^{\underline{h+1}}\right]|_{x=0} \\
&= -\sum_{j=1}^{h+1} \Delta^j \left[x^{\underline{h+1}}\right]|_{x=0}((1+\varepsilon)o_e)^j \\
&= -\sum_{j=1}^{h+1} \mathcal{S}(h+1, j) j!((1+\varepsilon)o_e)^j,
\end{aligned}$$

which completes the proof of the lemma. In particular, the first equality follows from the fact that $\Delta^{h+1}\left[k_e^{\underline{h+1}}\right]$ is constant with respect to k_e, so that its value does not change if we set $k_e = 0$. The last equality comes from the properties of the forward finite differences described in Section 8.2. □

Because of Lemma 9.2, and since $\sum_{j=1}^{h+1}\mathcal{S}(h+1, j)j!((1+\varepsilon)o_e)^j \leq o_e^{h+1}\mathcal{E}_{h+1}(1+\varepsilon)$ and $\mathcal{E}_{h+1}(1+\varepsilon) \leq \mathcal{E}_{d+1}(1+\varepsilon)$ for each $h \in [d]_0$, we have that the dual constraint of DLP is satisfied. Thus the considered tax function is $\mathcal{E}_{d+1}(1+\varepsilon)$-efficient.

By exploiting the same arguments used in the proof of Theorem 9.2, one can easily obtain that the tax $\delta \in \mathsf{TAX}_{\mathsf{MSP}}(\mathsf{U}(\mathcal{P}(d)))$ based on a mixed strategy profile $\overline{\mathbf{y}} \in \boldsymbol{\Delta}$ approximating the minimum of the (convex) function

$$\sum_{e\in E}\sum_{h\in[d]_0} \alpha_{e,h}\mathcal{E}_{h+1}((1+\varepsilon)\mathbb{E}[k_e(\mathbf{y})])$$

up to a factor $(1+\xi)$, is a $(1+\xi)\mathcal{E}_{d+1}(1+\varepsilon)$-efficient refundable tax. □

We now exhibit a matching lower bound holding for dynamic tax functions based on a social optimum.

Theorem 9.5. *Fix an integer $d \geq 1$. For any fixed $\xi > 0$, there exists $\mathsf{CG} \in \mathsf{U}(\mathcal{P}(d))$ such that for any $\delta \in \mathsf{TAX}_{\mathsf{OPT}}(\mathsf{W}(\mathcal{P}(d)))$, δ is not a $(\mathcal{E}_{d+1}(1+\varepsilon)-\xi)$-efficient refundable tax for CG with respect to ε-approximate one-round walks starting from the empty state and to the total latency.*

Proof. Fix an integer $d \geq 1$, a dynamic tax functions $\delta \in \mathsf{TAX}_{\mathsf{OPT}}(\mathsf{W}(\mathcal{P}(d)))$ and a value $\xi > 0$. Game CG has n players and a set of n resources $E = \{e_1, \ldots, e_n\}$. The latency function of resource j is $\ell_j(x) = \alpha_j x^d$ with $\alpha_j = \left(\frac{1+\varepsilon}{2+\varepsilon}\right)^{j-1}$. For each $i \in [n]$, $\Sigma_i = \{S_{i,1}, S_{i,2}\}$ with $S_{i,1} := \bigcup_{j=i+1}^{n}\{e_j\}$ and $S_{i,2} := \{e_i\}$ so that player n has one strategy only.

First, we show that the unique social optimum with respect to the total latency is given by the strategy profile $\boldsymbol{\sigma}^* := (S_{1,2}, S_{2,2}, \ldots, S_{n,2})$. To this aim consider a strategy profile $\boldsymbol{\sigma} \neq \boldsymbol{\sigma}^*$ and let i be the first player choosing her first strategy. Let us evaluate the difference $\mathsf{TL}(\boldsymbol{\sigma}) - \mathsf{TL}(\boldsymbol{\sigma}_{-i}, S_{i,2})$. We distinguish between two cases. If $\sigma_{i+1} = S_{i+1,2}$, we get $\mathsf{TL}(\boldsymbol{\sigma}) - \mathsf{TL}(\boldsymbol{\sigma}_{-i}, S_{i,2}) \geq 3\alpha_{i+1} - \alpha_i > 0$. If otherwise, $\sigma_{i+1} = S_{i+1,1}$ then we have $i+1 < n$ so that $\mathsf{TL}(\boldsymbol{\sigma}) - \mathsf{TL}(\boldsymbol{\sigma}_{-i}, S_{i,2}) \geq \alpha_{i+1} + 3\alpha_{i+2} - \alpha_i > 0$. Hence, the unique social optimum is $\boldsymbol{\sigma}^*$.

Consider now the strategy profile $\boldsymbol{\sigma} = (S_{1,1}, S_{2,1}, \ldots, S_{n-1,1}, S_{n,2})$. We show that there exists an ε-approximate one-round walk starting from the empty state $\boldsymbol{\tau} = (\boldsymbol{\sigma}^0, \boldsymbol{\sigma}^1, \ldots, \boldsymbol{\sigma}^n)$ such that $\boldsymbol{\sigma}$ is the strategy profile generated by $\boldsymbol{\tau}$. For a generic player $i \in N \setminus \{n\}$, we have

$$\begin{aligned}
\widehat{c}_i(\boldsymbol{\sigma}^{i-1}_{-i}, S_{i,1}) &= \sum_{j=i+1}^{n} \left(\frac{1+\varepsilon}{2+\varepsilon}\right)^{j-1} \left(i^d + \delta_d(1,i)\right) \\
&\leq \sum_{j=i+1}^{\infty} \left(\frac{1+\varepsilon}{2+\varepsilon}\right)^{j-1} \left(i^d + \delta_d(1,i)\right) \\
&= (i^d + \delta_d(1,i)) \left(\frac{1+\varepsilon}{2+\varepsilon}\right)^{i} \sum_{j=0}^{\infty} \left(\frac{1+\varepsilon}{2+\varepsilon}\right)^{j} \\
&= (i^d + \delta_d(1,i)) \left(\frac{1+\varepsilon}{2+\varepsilon}\right)^{i} (2+\varepsilon) \\
&= (i^d + \delta_d(1,i))(1+\varepsilon) \left(\frac{1+\varepsilon}{2+\varepsilon}\right)^{i-1}
\end{aligned}$$

whereas

$$\widehat{c}_i(\boldsymbol{w}^{i-1}_{-i}, S_{i,2}) = (i^d + \delta_d(1,i)) \left(\frac{1+\varepsilon}{2+\varepsilon}\right)^{i-1}.$$

Thus, we can conclude that $\boldsymbol{\sigma}^n = \boldsymbol{\sigma}$.

Let us now evaluate the ratio $\mathsf{TL}(\boldsymbol{\sigma})/\mathsf{TL}(\boldsymbol{\sigma}^*)$ to obtain a lower bound on the efficiency of δ. Posing $x := \frac{1+\varepsilon}{2+\varepsilon}$ so that $\frac{1}{1-x} = 2+\varepsilon$ and $\frac{x}{1-x} = 1+\varepsilon$, we get

$$\begin{aligned}
\lim_{n\to\infty} \frac{\mathsf{TL}(\boldsymbol{\sigma})}{\mathsf{TL}(\boldsymbol{\sigma}^*)} &= \frac{\sum_{i=1}^{\infty} (i-1)^{d+1} x^{i-1}}{\sum_{i=1}^{\infty} x^{i-1}} = \frac{\sum_{j=0}^{\infty} j^{d+1} x^j}{\sum_{j=0}^{\infty} x^j} = \frac{\left(x\frac{\partial}{\partial x}\right)^{d+1} \sum_{j=0}^{\infty} x^j}{\sum_{j=0}^{\infty} x^j} \\
&= (1-x) \left(x\frac{\partial}{\partial x}\right)^{d+1} \frac{1}{1-x} = \mathcal{E}_{d+1}\left(\frac{x}{1-x}\right) = \mathcal{E}_{d+1}(1+\varepsilon),
\end{aligned}$$

where the second last equality follows from Euler's formula (8.4). □

For weighted congestion games, we can prove the following result.

Theorem 9.6. *Fix an integer $d \geq 1$. For any $\mathsf{CG} \in \mathsf{W}(\mathcal{P}(d))$, there exists $\delta \in \mathsf{TAX_{OPT}}(\mathsf{W}(\mathcal{P}(d)))$ such that δ is a $\mathcal{Z}_{d+1}(1+\varepsilon)$-efficient refundable tax with respect to ε-approximate one-round walks starting from the empty state and to the weighted total latency.*

Proof. For an integer $d \geq 1$, fix a weighted congestion game $\mathsf{CG} \in \mathsf{W}(\mathcal{P}(d))$ and let $\boldsymbol{\sigma}^*$ be a social optimum for CG with respect to the social function WTL. Consider a dynamic tax function $\delta \in \mathsf{TAX_{OPT}}(\mathsf{W}(\mathcal{P}(d)))$ based on $\boldsymbol{\sigma}^*$ such that

$$\delta_h(w, o_e, k_e) := \delta_h(o_e, k_e) := \sum_{i=0}^{h} (h+1)_{i+1}((1+\varepsilon)o_e)^i(k_e + io_e)^{h-i} - k_e^h.$$

First of all, we show that $\delta_h(o_e, k_e) \geq 0$ for each pair of integers (o_e, k_e) such that $o_e \geq 0$ and $k_e \geq 1$ and that $\delta_h(o_e, k_e)$ is non-decreasing in k_e, so that δ is indeed a tax function and the inequalities derived in Section 9.1 can be applied. We get

$$\begin{aligned}\delta_h(o_e, k_e) :=& \sum_{i=0}^{h} (h+1)_{i+1}((1+\varepsilon)o_e)^i(k_e + io_e)^{h-i} - k_e^h \\ =& (h+1)k_e^h - k_e^h + \sum_{i=1}^{h} (h+1)_{i+1}((1+\varepsilon)o_e)^i(k_e + io_e)^{h-i}.\end{aligned}$$

Since $(h+1)k_e^h - k_e^h = hk_e^h$ and $\sum_{i=1}^{h}(h+1)_{i+1}((1+\varepsilon)o_e)^i(k_e+io_e)^{h-i}$ are polynomials with non-negative coefficients, the claim holds.

Now let $\boldsymbol{\sigma}$ be the strategy profile generated by an ε-approximate one-round walk starting from the empty state for (CG, δ). By applying the primal-dual method, we get the linear program $\mathsf{LP} := \mathsf{LP}(\mathsf{WTL}, \boldsymbol{\sigma}, \boldsymbol{\sigma}^*, \delta)$ with *constraints*$(\boldsymbol{\sigma}, \boldsymbol{\sigma}^*)$ equal to inequality (9.5) (see Section 9.1). The related dual program DLP is the following one (we associate the dual variables x and γ with the first and second constraint of LP, respectively):

$$\begin{aligned}&\min \gamma \\ &s.t.\ \ k_e^{h+1} \leq \gamma o_e^{h+1} + x\left(\int_0^{k_e}(t^h + \delta_h(o_e,t))dt - o_e((k_e+o_e)^h + \delta_h(o_e, k_e+o_e))\right) \\ &\qquad x \geq 0.\end{aligned}$$

Now we show that, for each $e \in E$ and $h \in [d]_0$, the dual constraint obtained by setting $x = 1$ and $\gamma = \mathcal{Z}_{d+1}(1+\varepsilon)$ is always satisfied by the proposed tax function δ. To this aim, we exploit the following result.

Lemma 9.3. *For each $e \in E$ and $h \in [d]_0$, $-k_e^{h+1} + \int_0^{k_e}\left(t^h + \delta_h(o_e,t)\right)dt - (1+\varepsilon)o_e\left((k_e+o_e)^h + \delta_h(o_e, k_e+o_e)\right) = -\mathcal{Z}_{h+1}(1+\varepsilon)o_e^{h+1}$.*

Proof (of the lemma). Define

$$T_{h,i}(o_e, k_e) := \sum_{r=i}^{h} (h+1)_r(h+1-r)((1+\varepsilon)o_e)^r(k_e + ro_e)^{h-r}$$

so that, for each $i \in [h+1]_0$,

$$T_{h,0}(o_e,k_e) = \delta_h(o_e,k_e) + k_e^h, \tag{9.14}$$

$$T_{h,i}(o_e,k_e) = (h+1)_{i+1}((1+\varepsilon)o_e)^i(k_e+io_e)^{h-i} + T_{h,i+1}(o_e,k_e). \tag{9.15}$$

One can prove by induction on $i \in [h+1]_0$ that

$$\begin{aligned}
&-k_e^{h+1} + \int_0^{k_e}(t^h+\delta_h(o_e,t)) - (1+\varepsilon)o_e\Big((k_e+o_e)^h + \delta_h(o_e,k_e+o_e)\Big)\\
&= -(h+1)_i((1+\varepsilon)o_e)^i(k_e+io_e)^{h+1-i} - \sum_{j=0}^{i-1}(1+\varepsilon)^j o_e^{h+1} j^{h+1-j}(h+1)_j\\
&\qquad + \int_0^{k_e} T_{h,i}(o_e,t)dt - (1+\varepsilon)o_e T_{h,i}(o_e,k_e+o_e). \qquad (9.16)
\end{aligned}$$

Then, by using $i = h+1$ in (9.16), we get

$$\begin{aligned}
&-k_e^{h+1} + \int_0^{k_e}(t^h+\delta_h(o_e,t)) - (1+\varepsilon)o_e\Big((k_e+o_e)^h + \delta_h(o_e,k_e+o_e)\Big)\\
&= -(h+1)_{h+1}((1+\varepsilon)o_e)^{h+1} - \sum_{j=0}^{h+1-1}(1+\varepsilon)^j o_e^{h+1} j^{h+1-j}(h+1)_j\\
&= -\sum_{j=0}^{h+1}(1+\varepsilon)^j o_e^{h+1} j^{h+1-j}(h+1)_j = -\mathcal{Z}_{h+1}(1+\varepsilon)o_e^{h+1},
\end{aligned}$$

which completes the proof of the lemma. □

Finally, the claim of the theorem follows since $\mathcal{Z}_{h+1}(1+\varepsilon) \leq \mathcal{Z}_{d+1}(1+\varepsilon)$ for each $h \in [d]_0$. □

9.3 Refundable Taxes for Non-atomic One-Round Walks

Recall definitions and notations on non-atomic one-round walks given in Chapter 7. In the following theorem, we prove that, by resorting to refundable dynamic taxes, the ε-competitive ratio in non-atomic congestion games with polynomial latency functions can be consistently lowered.

Theorem 9.7. *Fix an integer $d \geq 1$. For any non-atomic congestion game $\mathsf{NCG} \in \mathsf{N}(\mathcal{P}(d))$, there exists $\delta \in \mathsf{TAX}_{\mathsf{OPT}}(\mathsf{N}(\mathcal{P}(d)))$ such that δ is a $(1+\varepsilon)^{d+1}(d+1)!$-efficient refundable tax with respect to ε-approximate non-atomic one-round walks starting from the empty state.*

Proof. For an integer $d \geq 1$, fix a non-atomic congestion game $\mathsf{NCG} \in \mathsf{N}(\mathcal{P}(d))$, and consider the corresponding game (NCG, δ) with taxes, where δ is a dynamic tax-function defined as follows:

$$\delta_e(k_e) := \sum_{h=0}^{d} \alpha_{e,h}\delta_{e,h}(k_e) \quad \text{with} \quad \delta_{e,h}(k_e) := \sum_{j=1}^{h} (1+\varepsilon)^j (h)_j k_e^{h-j} o_e^j. \tag{9.17}$$

Let $\boldsymbol{\tau} = [(\boldsymbol{\sigma}^t)_{t\in[0,R]}, H]$ be an arbitrary ε-one-round walk of NCG, and let $\boldsymbol{\sigma}$ and $\boldsymbol{\sigma}^*$ be the strategy profile generated by $\boldsymbol{\tau}$ and the optimal one, respectively; let k_e and o_e denote $k_e(\boldsymbol{\sigma})$ and $k_e(\boldsymbol{\sigma}^*)$, respectively. As done in the proof of Theorem 7.1, by applying the primal-dual method, by relaxing/summing/integrating the constraints modeling ε-best-responses (which take into account taxation in this case, so that the latency functions perceived by each player are increased by δ_e), and by considering the related dual program, we get the following fact: given $\gamma, x \geq 0$ such that

$$-k_e^{h+1} + x\left(\frac{k_e^{h+1}}{h+1} + \int_0^{k_e} \delta_{e,h}(s)\mathrm{d}s - (1+\varepsilon)o_e(k_e^h + \delta_{e,h}(k))\right) + \gamma o_e^{h+1} \geq 0$$

for each $e \in E$ and $h \in [d]_0$, it follows that γ is an upper bound on the competitive ratio of NCG. We have that these constraints are always satisfied if $x := d+1$ and $\gamma := (1+\varepsilon)^{d+1}(d+1)!$. To show this fact, it suffices to prove the following inequality for each $h \in [d]_0$, $k \geq 0$ and $o \geq 0$:

$$\begin{aligned} &-k^{h+1} + (d+1)\left(\frac{k^{h+1}}{h+1} + \int_0^k \delta_h(s)\mathrm{d}s - (1+\varepsilon)o(k^h + \delta_h(k))\right) \\ &\quad + (1+\varepsilon)^{d+1}(d+1)!o^{h+1} \geq 0. \end{aligned}$$

Set $\delta_{h,l}(k) := \sum_{j=l}^{h}(1+\varepsilon)^j(h)_j o^j k^{h-j}$, so that $\delta_{h,l}(k) = (1+\varepsilon)^l(h)_l o^l k^{h-l} + \delta_{h,l+1}(k)$. First of all, one can prove by induction on $l \in [h+1]$ that the following equality holds:

$$\begin{aligned} &\int_0^k \delta_h(s)\mathrm{d}s - (1+\varepsilon)o(k^h + \delta_h(k)) \\ &= \int_0^k \delta_{h,l}(s)\mathrm{d}s - (1+\varepsilon)^l(h)_{l-1}o^l k^{h+1-l} - (1+\varepsilon)o\delta_{h,l}(k). \end{aligned} \tag{9.18}$$

Then, by setting $l := h+1$ in (9.18), we get:

$$\int_0^k \delta_h(s)\mathrm{d}s - (1+\varepsilon)o(k^h + \delta_h(k)) = -(1+\varepsilon)^{h+1}h!o^{h+1}. \tag{9.19}$$

We have

$$\begin{aligned} &-k^{h+1} + (d+1)\left(\frac{k^{h+1}}{h+1} + \int_0^k \delta_h(s)\mathrm{d}s - (1+\varepsilon)ok^h - (1+\varepsilon)\delta_h(k)\right) + \\ &\quad + (1+\varepsilon)^{d+1}(d+1)!o^{h+1} \\ &= -k^{h+1} + (d+1)\left(\frac{k^{h+1}}{h+1} - (1+\varepsilon)^{h+1}h!o^{h+1}\right) + (1+\varepsilon)^{d+1}(d+1)!o^{h+1} \end{aligned} \tag{9.20}$$

$$\geq -k^{h+1} + (d+1)\left(\frac{k^{h+1}}{d+1} - (1+\varepsilon)^{h+1}h!o^{h+1}\right) + (1+\varepsilon)^{d+1}(d+1)!o^{h+1}$$
$$= -(1+\varepsilon)^{h+1}(d+1)h!o^{h+1} + (1+\varepsilon)^{d+1}(d+1)d!o^{h+1} \geq 0,$$

where equality (9.20) comes from (9.19). □

Remark 9.3. Observe that the problem of finding an optimal strategy profile requires to solve a convex minimization problem with linear constraints and a polynomial objective function. This problem can be approximately solved in polynomial time by resorting to interior-point methods (see Boyd and Vandenberghe [43]). Therefore, for each $\xi > 0$, one can compute in polynomial time a strategy profile whose social value is at most $1+\xi$ times the social optimum. Then, if $\tilde{o}_e$ is the congestion of resource e in that strategy profile, by using the tax function defined in (9.17) with $\tilde{o}_e$ in place of o_e, one can easily prove that the resulting ε-competitive ratio is at most $(1+\xi)$ times that of Theorem 9.7.

9.4 Static versus Dynamic Taxes

Together with the dynamic taxes considered above, we can design efficient static tax functions based on optimal strategy profiles, for the case of polynomial latencies. Such taxes can be defined as $\delta_e(k_e) := \sum_{h\in[d]_0} \alpha_{h,e} \cdot \delta(o_e)$, where o_e, as usual, denotes the optimal congestion of resource e. By exploiting similar arguments as in Chapter 4, we can guarantee the existence of $(1+\varepsilon)\phi^d_{\varepsilon,d+1}$-efficient static taxes based on optimal strategy profiles in weighted (resp. weighted, resp. non-atomic) congestion games with polynomial latency functions of maximum degree d, where $\phi_{\varepsilon,d+1}$ is the unique positive solution k of equation $g(k) := -k^{d+1} + k + (1+\varepsilon)((k+1)^d - 1) = 0$ (resp. $g(k) := -\frac{k^{d+1}}{d+1} + k + (1+\varepsilon)((k+1)^d - 1) = 0$, resp. $k := (1+\varepsilon)(d+1)$) with respect to ε-approximate pure Nash equilibria (resp. ε-approximate one-round walks, resp. ε-approximate non-atomic one round walks). For unweighted games, we have obtained static tax functions based on optimal strategy profiles (with respect to both ε-approximate pure Nash equilibria or one-round walks) whose efficiency is as high as that in weighted games if $\phi_{\varepsilon,d+1}$ is integer, and is equal to $h(\lfloor\phi_{\varepsilon,d+1}\rfloor)$, where $h(k) = kf(k) + xg(k)$ and $x \geq 0$ is such that $h(\lfloor\phi_{\varepsilon,d+1}\rfloor) = h(\lfloor\phi_{\varepsilon,d+1} + 1\rfloor)$.

Furthermore, by exploiting similar techniques as in Chapter 4 to construct tight lower bounds, one can show that the obtained efficiency is the best possible under static taxes based on optimal strategy profiles. This fact reveals that static taxes based on optimal strategy profiles are better than the case without taxes, but worse than dynamic taxes. In Figure 9.2, we show some numerical comparison for the case of pure Nash equilibria in unweighted congestion games.

Degree	No Taxes	Static Taxes	Dynamic Taxes
1	2.5	2	2
2	9.583	5.25	5
3	41.54	18.36	15
4	267.6	89.41	52
5	1,514	469.7	203
6	12,345	3,325	877
7	98,734	24,070	4,140
8	802,603	185,807	21,147
9	10,540,286	2,101,636	115,975
10	88,562,706	17,275,286	678,570

Fig. 9.2 Comparison of the efficiency obtainable in unweighted congestion games with polynomial latency functions with maximum degree $d \in [10]$ and with respect to pure Nash equilibria, in three different settings: without taxes, with static taxes and general dynamic taxes based on optimal strategy profiles. Observe that the more powerful the tax function, the better the efficiency.

9.5 Non-Refundable Taxes

In this section, we give a brief overview of the results obtained for non-refundable taxes. By exploiting similar arguments to those used for refundable taxes, one can show how to define efficient dynamic tax functions based on a social optimum with respect to both ε-approximate pure Nash equilibria and ε-approximate one-round walks starting from the empty state.

Theorem 9.8. *Fix an integer* $d \geq 1$. *For any* $\mathsf{CG} \in \mathsf{W}(\mathcal{P}(d))$, *there exists* $\delta \in \mathsf{TAX}_{\mathsf{OPT}}(\mathsf{W}(\mathcal{P}(d)))$ *such that* δ *is a* $\left(\min_{x>1}\left\{(1+\varepsilon)x\mathcal{R}_d\left(\frac{(1+\varepsilon)x}{x-1}\right)\right\}\right)$*-efficient non-refundable tax with respect to* ε*-approximate pure Nash equilibria and to the total latency.*

Theorem 9.9. *Fix an integer* $d \geq 1$. *For any* $\mathsf{CG} \in \mathsf{W}(\mathcal{P}(d))$, *there exists* $\delta \in \mathsf{TAX}_{\mathsf{OPT}}(\mathsf{W}(\mathcal{P}(d)))$ *such that* δ *is a* $\mathcal{Q}_{d+1}(1+\varepsilon)$*-efficient non-refundable tax with respect to* ε*-approximate one-round walks starting from the empty state and to the weighted total latency.*

The proofs of the above theorems are quite technical and have been omitted. To show the above theorems, as in the refundable case, we start from a linear program $\mathsf{LP}(\mathsf{SF}, \boldsymbol{\sigma}, \boldsymbol{\sigma}^*)$ as that defined in Section 9.1, but with $\mathsf{SF} \in \{\mathsf{TC}, \mathsf{WTC}\}$ and new primal constraints, and then we construct the dual in variables (x, γ) whose objective function is equal to γ. Finally, the taxes are determined in such a way that the feasible value γ is the lowest possible.

9.6 Concluding Remarks and Related Work

The content of this chapter is based on the work of Bilò and Vinci [27], that generalizes the work of Caragiannis et al. [52] from affine to polynomial latency functions, and considers other solution concepts (e.g., approximate equilibria and one-round walks). Among the obtained results, Caragiannis et al. [52] provides a 2-upper bound for refundable taxes with respect to mixed Nash equilibria and to both the total latency and the weighted total latency, and also obtains a 4-upper bound for non-refundable taxes with respect to mixed Nash equilibria and to the total latency. We observe that such upper bounds can be reobtained as a special case of our bounds by setting $\varepsilon = 0$ and $d = 1$. In these cases, our taxes are indeed static ones.

For further related works, the interested reader can refer to the concluding sections of Chapters 8 and 10.

Chapter 10
Dynamic Taxes for Unweighted Congestion Games: from Polynomial to General Latencies

In this chapter, we focus on the problem of defining refundable dynamic taxes for unweighted congestion games with general latency functions, thus generalizing part of Chapter 9 (in which we have focused on the specific case of polynomial latency functions), and we give an almost complete characterization of the efficiency of taxation in unweighted congestion games, as done by Pigou [150] and Beckmann et al. [17] for non-atomic congestion games.

Under mild assumptions on the considered latency functions, we explicitly define and analyze efficient refundable dynamic taxes for unweighted congestion games with general latency functions, and with respect to both the ϵ-approximate price of anarchy and the competitive ratio of ϵ-approximate one-round walks. Furthermore, we show that, under mild assumptions on the considered classes of taxes, the efficiency of our taxes is the best possible, even for load balancing games, and for load balancing games with identical resources if $\epsilon = 0$.

Similar results are given for non-atomic congestion games too, where efficient taxes have been designed with the aim of reducing the competitive ratio of ϵ-approximate non-atomic one-round walks.

10.1 Refundable Taxes for Unweighted Games and Pure Nash Equilibria

In this section, we show how to define an efficient dynamic tax function based on a social optimum with respect to ϵ-approximate pure Nash equilibria for unweighted games with strictly semi-convex latency functions. Before providing the main results, we first show a preliminary lemma based on the application of the primal-dual method.

Lemma 10.1. *Let $\mathcal{G}$ be a class of unweighted congestion games, $\mathsf{CG} \in \mathcal{G}$ be a fixed congestion game, and $\delta \in \mathsf{TAX}_{\mathsf{OPT}}(\mathcal{G})$ be a refundable tax for CG defined as $\delta_e(k) := \delta(k_e, o_e, \ell_e)$, where o_e and ℓ_e denote the optimal congestion and the latency*

V. Bilò, C. Vinci, *Coping with Selfishness in Congestion Games*, Monographs in Theoretical Computer Science. An EATCS Series, https://doi.org/10.1007/978-3-031-30261-9_10

function of the considered resource e, respectively. We have that δ is γ^*-efficient for CG *if $\delta(k,o,f)$ is non-decreasing in k for any fixed o, f, and the following inequalities are satisfied for any $k,o \in \mathbb{Z}_{\geq 0}$ and $f \in \mathcal{C}(\mathcal{G})$:*

$$\gamma^* of(o) \geq -k\delta(k,o,f) + (1+\varepsilon)of(k+1) + (1+\varepsilon)o\delta(k+1,o,f). \tag{10.1}$$

Proof. Let $\boldsymbol{\sigma}$ and $\boldsymbol{\sigma}^*$ be an ε-approximate equilibrium in game (CG, δ) and an optimal strategy profile of CG, respectively. By applying the primal-dual method, and by using similar arguments as in Section 9.1 (for the use of taxes) and Chapter 4 (to cope with general latency functions), we have that the optimal value of the following linear program LP in the variables α_es is an upper bound on the efficiency of tax function δ with respect to ε-approximate pure Nash equilibria:

$$\begin{aligned}
\mathsf{LP}: \quad & \max \sum_{e\in E} \alpha_e k_e \ell_e(k_e) \\
& s.t. \sum_{e\in E} \alpha_e \left[-k_e(\ell_e(k_e) + \delta(k_e,o_e,\ell_e))\right] \\
& \qquad + \sum_{e\in E} \alpha_e \left[(1+\varepsilon)o_e(\ell_e(k_e+1) + \delta(k_e+1,o_e,\ell_e))\right] \geq 0 \qquad (10.2) \\
& \sum_{e\in E} \alpha_e o_e \ell_e(o_e) = 1 \\
& \alpha_e \geq 0, \quad \forall e \in E,
\end{aligned}$$

where (10.2) is a relaxed Nash equilibrium condition coming from the fact that $\delta(k,o,f)$ is non-decreasing in k. The dual program DLP (in dual variables x, γ) is the following:

$$\begin{aligned}
\mathsf{DLP}: \quad & \min \quad \gamma \\
& s.t. \quad \gamma o_e \ell_e(o_e) \geq k_e \ell_e(k_e) + x\left[-k_e(\ell_e(k_e) + \delta(k_e,o_e,\ell_e))\right] \\
& \qquad + x\left[(1+\varepsilon)o_e(\ell_e(k_e+1) + \delta(k_e+1,o_e,\ell_e))\right], \ \forall e \in E \qquad (10.3) \\
& \qquad x \geq 0.
\end{aligned}$$

Any feasible solution γ, by the Weak Duality Theorem, is an upper bound on the efficiency of tax function δ. After setting $x = 1$ in (10.3), we get the following equivalent set of constraints in variable γ:

$$\gamma o_e \ell_e(o_e) \geq -k_e\delta(k_e,o_e,\ell_e) + (1+\varepsilon)o_e(\ell_e(k_e+1) + \delta_e(k_e+1,o_e,\ell_e)), \ \forall e \in E. \tag{10.4}$$

By exploiting (10.1), we have that (10.4) is feasible if we set $\gamma := \gamma^*$, and this shows the claim. □

Given $o \in \mathbb{N}$, a class of latency functions $\mathcal{G}$, and $f \in \mathcal{G}$, let:

$$\mathcal{B}_f(1+\varepsilon,o) := \frac{\sum_{j=0}^{\infty} \frac{(1+\varepsilon)^j o^j}{j!} jf(j)}{\sum_{j=0}^{\infty} \frac{(1+\varepsilon)^j o^j}{j!} of(o)} = \frac{\sum_{j=0}^{\infty} \frac{(1+\varepsilon)^j o^j}{j!} jf(j)}{e^{(1+\varepsilon)o} of(o)}, \tag{10.5}$$

$$\mathcal{B}_{\mathcal{G}}(1+\varepsilon) := \sup_{f \in \mathcal{C}(\mathcal{G}), o \in \mathbb{N}} \mathcal{B}_f(1+\varepsilon, o). \tag{10.6}$$

By applying Lemma 10.1, one can show the following theorem.

Theorem 10.1. *Let $\mathcal{G}$ be a class of unweighted games whose non-constant latency functions are strictly semi-convex. Given $\mathsf{CG} \in \mathcal{G}$, there exists a $\mathcal{B}_{\mathcal{G}}(1+\varepsilon)$-efficient refundable tax function $\delta \in \mathsf{TAX}_{\mathsf{OPT}}(\mathcal{G})$ for CG, with respect to ε-approximate pure Nash equilibria and to the total latency.*

Proof (Sketch). For any fixed non-constant latency $f \in \mathcal{C}(\mathcal{G})$ and $o \in \mathbb{N}$, let $\delta(k,o,f)$ be the tax function inductively defined as follows: (i) we set $\delta(k,o,f) := 0$ for $k := 0$; (ii) we impose that (10.1) holds as equality for the considered o, f and γ^* equal to $\gamma^*_{o,f} := \mathcal{B}_f(1+\varepsilon, o)$, so that we can iteratively define $\delta(k,o,f)$ for any $k > 0$. If f is constant or $o = 0$, we set $\delta(k,o,f) := 0$ for any k.

If f is constant or $o = 0$, we have that (10.1) is trivially verified if $\gamma^* = 1$, and then it is also satisfied by $\gamma^* := \mathcal{B}_{\mathcal{G}}(1+\varepsilon) \geq 1$. Now, assume that f is non-constant and $o > 0$. By the hypothesis, we necessarily have that f is strictly semi-convex. By explicating the recurrence relation to compute $\delta(k,o,f)$ and by exploiting the definition of $\mathcal{B}_f(1+\varepsilon, o)$, one can show that $\delta(k,o,f)$ is non-negative for any integer $k \geq 0$. By using the strict semi-convexity of f and the fact that (10.1) is satisfied as equality by $\delta(k,o,f)$ for any fixed o if $\gamma^* = \gamma^*_{o,f}$, one can show that $\delta(k,o,f)$ is necessarily non-decreasing in k. Thus, by setting $\gamma^* := \mathcal{B}_{\mathcal{G}}(1+\varepsilon)$ in (10.1) and by exploiting the above properties of tax function $\delta(k,o,f)$, we have that the set of constraints (10.1) is satisfied if f is non-constant and $o > 0$ (since $\mathcal{B}_{\mathcal{G}}(1+\varepsilon) \geq \gamma^*_{o,f}$ in this case).

We conclude that our tax function satisfies the hypothesis of Lemma 10.1, and this shows the claim. □

Remark 10.1. As (non-constant) polynomial latency functions are strictly semi-convex, we can apply Theorem 10.1 to reobtain automatically the claim of Theorem 9.1, for the restricted case of unweighted games.

Remark 10.2. By exploiting the convexity of the latency functions, the arguments of Theorem 9.2 on how to compute efficient taxes in polynomial time can be easily extended to more general latency functions. In particular, one can show that, for any $\xi > 0$, $(\mathcal{B}_{\mathcal{G}}(1+\varepsilon)+\xi)$-efficient taxes can be computed by solving a convex optimization problem, whose $(1+\xi)$-approximate optimum can be found in polynomial time under mild assumptions on the latency functions.

Now, we exhibit a matching lower bound holding for dynamic tax functions based on a social optimum and with strictly semi-convex latency functions, which becomes a corollary of the following theorem.

Theorem 10.2. *Let $\mathcal{G}$ be a class of unweighted congestion games whose non-constant latency functions are closed under ordinate scaling and are strictly semi-convex. For any fixed $M < \mathcal{B}_{\mathcal{G}}(1+\varepsilon)$, there exists $\mathsf{CG} \in \mathcal{G}$ such that, for any*

$\delta \in \mathsf{TAX}_{\mathsf{OPT}}(\mathcal{G})$, δ is not an M-efficient refundable tax for CG *with respect to ε-approximate pure Nash equilibria and to the total latency. Furthermore, the considered lower bound holds even for load balancing games*[1]*, and for load balancing games with identical resources if $\varepsilon = 0$.*

Proof (Sketch). Let $f \in \mathcal{C}(\mathcal{G})$ and $o \in \mathbb{N}$ be such that $\mathcal{B}_f(1+\varepsilon, o) > M$, and fix an arbitrary integer $n > o$. Similarly as in Theorems 4.2 and 4.3, we define the lower bounding instance by resorting to a *load balancing (multi-)graph*: the nodes are the resources, and each copy of an edge (u,v) corresponds to a player who can play resources u,v only, denoted, respectively, as the first and the second strategy of the player.

Consider a load balancing graph $\mathsf{LB}(n,o)$ organized in $n+1$ levels numbered from 0 to n and defined as follows. For each $j \in [n]$, there are $o^{n-j}(n-o)(n-1)_{j-1}$ nodes/resources at level j, and there are o^n nodes/resources at level 0. There are o self-loop edges on each resource at level 0 (corresponding to dummy players with a unique strategy), and non-self loop edges can only connect nodes of consecutive levels, so that, for each $i \in [n]_0$, there are directed edges from level i to level $i+1$. The out-degree (including self-loop edges) of each node at level $j \in [n]_0$ is $n-j$ and the in-degree (including self-loop edges) of each node at level $j \in [n]_0$ is o. Each resource at level j has latency function $g_j(x) = f(x)/(1+\varepsilon)^j$. We observe that a load balancing graph satisfying the above properties can be realized.

Let $\boldsymbol{\sigma}$ and $\boldsymbol{\sigma}^*$ be the strategy profiles in which all players play their first and their second strategy, respectively. Observe that $k_e(\boldsymbol{\sigma}^*) = o$ for any $e \in E$ and $k_e(\boldsymbol{\sigma}) = n-j$ for any resource at level $j \in [n]_0$. See Figure 10.1 for a clarifying example on the considered load-balancing graph.

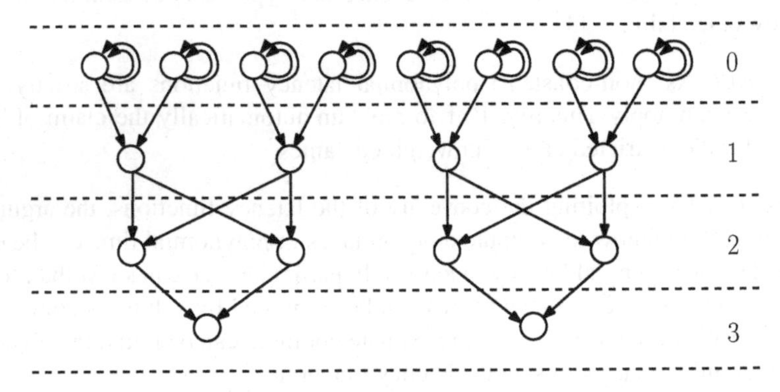

Fig. 10.1 The load balancing graph $\mathsf{LB}(n,o)$ described in Theorem 10.2, with $n = 3$ and $o = 2$. The out-degree and in-degree of each node/resource (including the contribution of self-loops) represents its congestion in $\boldsymbol{\sigma}$ and $\boldsymbol{\sigma}^*$, respectively. The numbers on the right denote the level of the corresponding resources on the left.

[1] As usual, the hypothesis of ordinate-scaling latencies is not necessary to provide lower bounds for general congestion games. However, in this case, it is necessary to show that the lower bound holds for general load balancing games.

By using the fact that latencies are strictly semi-convex, one can show the following lemma.

Lemma 10.2. *For any optimal strategy profile $\boldsymbol{\sigma}'$ of* LB, *we get* $k_e(\boldsymbol{\sigma}') = o$.

Because of Lemma 10.2, we can consider (without loss of generality) dynamic taxes based on strategy profile $\boldsymbol{\sigma}^*$ only.

Now, we show that $\boldsymbol{\sigma}$ is an ε-approximate pure Nash equilibrium, for any tax function δ based on the social optimum $\boldsymbol{\sigma}^*$, that is unique up to strategy profiles having the same congestion on each resource (by Lemma 10.2)[2]. Since the considered tax function is based on the social optimum, we have that δ_e is defined as $\delta_e(k) := \hat{\delta}(k,o,f)/(1+\varepsilon)^j$ for any resource e at level j, where $\hat{\delta}$ is a function that only depends on the congestion k of the resource e (with respect to the considered strategy profile), the optimal congestion $o = k_e(\boldsymbol{\sigma}^*)$ and the function f. For any non-self-loop player i, we get:

$$\begin{aligned}
\hat{c}_i(\boldsymbol{\sigma}) &= \frac{f(k_{\sigma_i}(\boldsymbol{\sigma})) + \hat{\delta}(k_{\sigma_i}(\boldsymbol{\sigma}),o,f)}{(1+\varepsilon)^j} \\
&= \frac{f(n-j) + \hat{\delta}(n-j,o,f)}{(1+\varepsilon)^j} \\
&= \frac{f(k_{\sigma_i^*}(\boldsymbol{\sigma})+1) + \hat{\delta}(k_{\sigma_i^*}(\boldsymbol{\sigma})+1,o,f)}{(1+\varepsilon)^j} \\
&= (1+\varepsilon)\frac{f(k_{\sigma_i^*}(\boldsymbol{\sigma}_{-i},\sigma_i^*)) + \hat{\delta}(k_{\sigma_i^*}(\boldsymbol{\sigma}_{-i},\sigma_i^*),o,f)}{(1+\varepsilon)^{j+1}} \\
&= (1+\varepsilon)\hat{c}_i(\boldsymbol{\sigma}_{-i},\sigma_i^*),
\end{aligned}$$

thus $\boldsymbol{\sigma}$ is an ε-approximate pure Nash equilibrium. We get

$$\mathsf{TL}(\boldsymbol{\sigma}) = o^n n f(n) + \sum_{j=1}^{n} o^{n-j}(n-o)(n-1)_{j-1}\frac{(n-j)f(n-j)}{(1+\varepsilon)^j},$$

$$\mathsf{TL}(\boldsymbol{\sigma}^*) = o^{n+1} f(o) + \sum_{j=1}^{n} o^{n-j}(n-o)(n-1)_{j-1}\frac{of(o)}{(1+\varepsilon)^j}.$$

Therefore, by simple calculations, one can show that

$$\lim_{n\to\infty}\frac{\mathsf{TL}(\boldsymbol{\sigma})}{\mathsf{TL}(\boldsymbol{\sigma}^*)} = \lim_{n\to\infty}\frac{\frac{(1+\varepsilon)^n\mathsf{TL}(\boldsymbol{\sigma})}{(n-o)(n-1)!}}{\frac{(1+\varepsilon)^n\mathsf{TL}(\boldsymbol{\sigma}^*)}{(n-o)(n-1)!}} = \lim_{n\to\infty}\frac{\sum_{j=0}^{n}\frac{(1+\varepsilon)^j o^j}{j!}jf(j)}{\sum_{j=0}^{n}\frac{(1+\varepsilon)^j o^j}{j!}of(o)} = \mathcal{B}_f(1+\varepsilon,o) > M.$$

We conclude that, for a sufficiently large n, any refundable tax function based on the social optimum is not M-efficient for $\mathsf{LB}(n,o)$, and this shows the claim. □

---

[2] In this case, one can also prove that $\boldsymbol{\sigma}^*$ is the unique strategy profile such that all the congestions are equal to o, but it is not necessary for our aims.

Remark 10.3. Similarly to the approach considered in Chapter 4, the lower bound of Theorem 10.2 can be derived by exploiting a modified linear program $\overline{\mathsf{LP}}$ obtained from the dual program DLP in the proof of Lemma 10.1, after transforming the taxes in further dual variables. In particular, by setting $y_{k,o,f} := x(\delta(k,o,f) + f(k))$ in DLP, and assuming without loss of generality that the set of congestions of type k_e and o_e appearing in (10.3) is the whole set $[n]_0$, we get the following dual program in variables γ and $(y_{k,o,f})_{k\in[n+1]_0,o\in[n]_0 f\in\mathcal{C}(\mathsf{CG})}$:

$\overline{\mathsf{DLP}}$:

$$\begin{aligned}\min\quad & \gamma \\ s.t.\quad & \gamma o f(o) \geq k f(k) - k y_{k,o,f} + (1+\varepsilon) o y_{k+1,o,f},\ \forall k,o \in [n]_0, f \in \mathcal{C}(\mathsf{CG}) \quad (10.7) \\ & y_{k,o,f} \geq 0,\ \forall k \in [n+1]_0, o \in [n]_0, f \in \mathcal{C}(\mathsf{CG}).\end{aligned}$$

Analogously to the proof of Theorem 10.1, the optimum of $\overline{\mathsf{DLP}}$ is achieved when the dual constraints (10.7) are tight for some $o \in [n]$ and $f \in \mathcal{C}(\mathsf{CG})$, and if $y_{0,o,f} = y_{n+1,o,f} = 0$. In particular, there are some $o \in [n]$ and $f \in \mathcal{C}(\mathsf{CG})$ such that the following dual program in variables γ and $(y_k)_{k\in[n]}$, with $y_k := y_{k,o,f}$, is equivalent to $\overline{\mathsf{DLP}}$:

$$\begin{aligned}\overline{\mathsf{DLP2}}:\quad \min\quad & \gamma \\ s.t.\quad & \gamma o f(o) = k f(k) - k y_k + (1+\varepsilon) o y_{k+1}, \quad \forall k \in [n]_0 \\ & y_k \geq 0,\ \forall k \in [n]_0,\end{aligned}$$

where y_0 and y_{n+1} are set equal to 0, and then are not considered as variables. The dual program of $\overline{\mathsf{DLP2}}$ is a linear program in variables $(\alpha_k)_{k\in[n]_0}$ defined as follows:

$$\begin{aligned}\overline{\mathsf{LP}}:\quad \max\quad & \sum_{k\in[n]_0} \alpha_k k f(k) \\ s.t.\quad & \alpha_k k \leq \alpha_{k-1}(1+\varepsilon) o, \quad \forall k \in [n] \qquad (10.8) \\ & \sum_{k\in[n]_0} \alpha_k o f(o) = 1 \\ & \alpha_k \geq 0, \quad \forall k \in [n]_0.\end{aligned}$$

One can show that the optimum of DLP2 is guaranteed when all the dual variables of DLP2 (but y_0 and y_{n+1}) are strictly positive. Thus, by the complementary slackness conditions, we have that all the constrains in (10.8) are tight.

The tight constraints of type $\alpha_k k = \alpha_{k-1}(1+\varepsilon)o$ suggest how to construct a load balancing instance serving as tight lower bound:

(i) the resources are organized in levels $0, 1, \ldots, n$, and each player is associated with a type $k \in [n]$;
(ii) each player of type k can select among a resource at level $n-k$ and a resource at level $n-k+1$, that we call respectively first and second strategy of the con-

sidered player; furthermore, let $\boldsymbol{\sigma}$ and $\boldsymbol{\sigma}^*$ denote the strategy profiles in which all players select their first and second strategy, respectively;

(iii) for any $k \in [n]$, the total cost of the players selecting a resource at level $n-k$ in $\boldsymbol{\sigma}$ is equal to $\alpha_k k f(k) c_k$ for some constant $c_k > 0$, and the sum of all their costs coming from unilateral deviations in favor of the second strategy (i.e., a sum of costs of type $c_i(\boldsymbol{\sigma}_{-i}, \sigma_i^*)$) is $\alpha_{k-1} o f(k) c_k$, and we have that $\alpha_k k f(k) c_k = (1+\varepsilon)\alpha_{k-1} o f(k) c_k$ because of the tight constraints.

In particular, property (iii) is satisfied by the following modeling choices:

(a) given $k \in [n]$, the number of resources at each level $n-k$ is proportional to α_k, and the number of players selecting each resource at level $n-k$ (resp. $n-k+1$) when playing the first (resp. second) strategy must be equal to k (resp. o);

(b) the cost of each player of type k selecting her first strategy (i.e., a resource at level $n-k$) in $\boldsymbol{\sigma}$ is exactly $(1+\varepsilon)$ times the cost she would incur if she unilaterally deviated in favor of her second strategy (i.e., a resource at level $n-k+1$), thus, playing the first strategy is a pure Nash equilibrium.

By imposing the above conditions, and by considering some further refinements, one can obtain a lower bounding instance that is structurally analogue to that considered in Theorem 10.2.

10.2 Refundable Taxes for One-Round Walks

10.2.1 Unweighted Games

Analogously to Theorem 10.1, we obtain efficient taxes with respect to ε-approximate one-round walks. We first provide a preliminary lemma.

Lemma 10.3. *Let $\mathcal{G}$ be a class of unweighted congestion games, $\mathsf{CG} \in \mathcal{G}$ be a fixed congestion game, and $\delta \in \mathsf{TAX}_{\mathsf{OPT}}(\mathcal{G})$ be a refundable tax for CG defined as $\delta_e(k_e) := \delta(k_e, o_e, f_e)$, where o_e and ℓ_e denote the optimal congestion and the latency function of the considered resource e, respectively. We have that δ is γ^*-efficient for CG if $\delta(k,o,f)$ is non-decreasing in k for any fixed o, f, and the following inequality is satisfied for any $k, o \in \mathbb{Z}_{\geq 0}$ and $f \in \mathcal{C}(\mathcal{G})$:*

$$\gamma^* o f(o) \geq k f(k) - \sum_{h=1}^{k} f(h) - ((1+\varepsilon)o+1)\overline{\delta}(k,o,f) + (1+\varepsilon)o\left(f(k+1) + \overline{\delta}(k+1,o,f) - \overline{\delta}(k,o,f)\right), \tag{10.9}$$

where $\overline{\delta}(k,o,f)$ is defined as $\sum_{h=1}^{k} \delta(h,o,f)$.

Proof (Sketch). Let $\boldsymbol{\sigma}$ (resp. $\boldsymbol{\sigma}^*$) be the strategy profile generated by an ε-one-round walk in game (CG, δ) (resp. an optimal strategy profile of CG). By applying the primal-dual method, and by using similar arguments as in Chapter 4 (for approximate one-round walks and general latency functions) and Lemma 10.1, we have that the optimal value of the following linear program in variables α_es is an upper bound on the efficiency of tax function δ with respect to ε-approximate one round walks:

$$\begin{aligned}
&\mathsf{LP}: \\
&\max \sum_{e\in E} \alpha_e k_e \ell_e(k_e) \\
&s.t. \sum_{e\in E} \alpha_e \sum_{h=1}^{k_e} (\ell_e(h) + \delta(h, o_e, f_e)) \\
&\qquad \leq \sum_{e\in E} \alpha_e (1+\varepsilon) o_e (\ell_e(k_e+1) + \delta(k_e+1, o_e, f_e)) \qquad (10.10) \\
&\sum_{e\in E} \alpha_e o_e \ell_e(o_e) = 1 \\
&\alpha_e \geq 0, \quad \forall e \in E,
\end{aligned}$$

where (10.10) holds since σ is the strategy profile generated by an ε-approximate one-round walk and since $\delta(k,o,f)$ is non-decreasing in k. The dual of LP (in variables x, γ) is the following:

$$\begin{aligned}
&\mathsf{DLP}: \\
&\min \quad \gamma \\
&s.t. \quad \gamma o_e \ell_e(o_e) \geq k_e \ell_e(k_e) + x \left[-\sum_{h=1}^{k_e} \ell_e(k_e) - \overbrace{\overline{\delta}(k_e, o_e, \ell_e)}^{=\sum_{h=1}^{k_e} \delta(h, o_e, f_e)} \right] \\
&\qquad + x \left[(1+\varepsilon) o_e \left(\ell_e(k_e+1) + \overbrace{\overline{\delta}(k_e+1, o_e, \ell_e) - \overline{\delta}(k_e, o_e, \ell_e)}^{=\delta(k_e+1, o_e, f_e)} \right) \right], \forall e \in E \\
&\qquad x \geq 0.
\end{aligned}$$

The claim follows by analyzing DLP as in Lemma 10.1. □

Given $o > 0$, a class of latency functions $\mathcal{G}$, and $f \in \mathcal{G}$, let:

$$\mathcal{E}_f(1+\varepsilon, o) := \frac{\sum_{j=0}^{\infty} \left(\frac{(1+\varepsilon)o}{(1+\varepsilon)o+1}\right)^j jf(j)}{\sum_{j=0}^{\infty} \left(\frac{(1+\varepsilon)o}{(1+\varepsilon)o+1}\right)^j of(o)} = \frac{\sum_{j=0}^{\infty} \left(\frac{(1+\varepsilon)o}{(1+\varepsilon)o+1}\right)^j jf(j)}{((1+\varepsilon)o+1)of(o)}, \tag{10.11}$$

$$\mathcal{E}_{\mathcal{G}}(1+\varepsilon) := \sup_{f\in\mathcal{C}(\mathcal{G}), o\in\mathbb{N}} \mathcal{E}_f(1+\varepsilon, o). \tag{10.12}$$

By using similar arguments as in the proof of Theorem 10.2, we can resort to Lemma 10.3 to show the following theorem.

Theorem 10.3. *Let $\mathcal{G}$ be a class of unweighted games whose non-constant latency functions are strictly semi-convex. Given $\mathsf{CG} \in \mathcal{G}$, there exists a $\mathcal{E}_{\mathcal{G}}(1+\varepsilon)$-efficient refundable tax function $\delta \in \mathsf{TAX}_{\mathsf{OPT}}(\mathcal{G})$ for CG, with respect to ε-approximate one-round walks and to the total latency.*

Proof (Sketch). For any fixed non-constant latency $f \in \mathcal{C}(\mathcal{G})$ and $o \in \mathbb{N}$, let $\delta(k,o,f)$ be the tax function inductively defined as follows: (i) we set $\delta(k,o,f) := 0$ for $k := 0$; (ii) we impose that (10.9) holds as equality for the considered o, f and γ^* equal to $\gamma^*_{o,f} := \mathcal{E}_f(1+\varepsilon,o)$, so that we can iteratively define $\delta(k,o,f)$ for any $k > 0$. If f is constant or $o = 0$, we set $\delta(k,o,f) := 0$ for any k.

By proceeding as in the proof of Theorem 10.2, one can show that δ is non-negative and non-decreasing. Furthermore, one can show that (10.9) is satisfied by δ. We conclude that the hypotheses of Lemma 10.3 are satisfied, and then δ is $\mathcal{E}_{\mathcal{G}}(1+\varepsilon)$-efficient refundable. □

Remark 10.4. As (non-constant) polynomial latency functions are strictly semi-convex, we can apply Theorem 10.3 to reobtain automatically the claim of Theorem 9.4.

Remark 10.5. By similar argument as in Remark 10.2, one can show that, for any $\xi > 0$, $(\mathcal{E}_{\mathcal{G}}(1+\varepsilon)+\xi)$-efficient taxes can be computed by solving a convex optimization problem, whose approximately optimal solution can be found in polynomial time under mild assumptions on the latency functions.

In Theorem 10.4, we exhibit a matching lower bound holding for dynamic tax functions based on a social optimum, thus generalizing Theorem 9.5 (holding for polynomial latencies only), that becomes a corollary of the following theorem.

Theorem 10.4. *Let $\mathcal{G}$ be a class of unweighted congestion games whose non-constant latency functions are closed under ordinate scaling and strictly semi-convex. For any fixed $M < \mathcal{E}_{\mathcal{G}}(1+\varepsilon)$, there exists $\mathsf{CG} \in \mathcal{G}$ such that, for any $\delta \in \mathsf{TAX}_{\mathsf{OPT}}(\mathcal{G})$, δ is not an M-efficient refundable tax for CG with respect to ε-approximate one-round walks and to the total latency. Furthermore, the considered lower bound holds even for load balancing games, and for load balancing games with identical resources if $\varepsilon = 0$.*

Proof (Sketch). Let $f \in \mathcal{C}(\mathcal{G})$ be a non-constant latency function and $o \in \mathbb{N}$ such that $\mathcal{E}_f(1+\varepsilon,o) > M$, and let $n > o$ be an arbitrary integer. We define the lower bounding instance by resorting to a load balancing graph, as in Theorem 10.2. Consider a load balancing graph $\mathsf{LB}(n,o)$ recursively defined as follows:

- the nodes/resources are split into different subgroups, and each subgroup is associated with a type $j \in [n]_0$ (in general, there are several subgroups associated with the same type);
- there is a unique subgroup of type n;

- for any subgroup of type $j \in [n]_0$, there are (directed) edges from this group to j subgroups of type $j-1, j-2, \ldots, 1, 0$;
- the subgraph induced by subgroups (i.e. a graph whose nodes are subgroups, and there is an edge (u,v) between two subgroups if and only if there are edges going from u to v in the load balancing graph) is a directed tree, whose edges are oriented from the root to the leaves;
- each resource belonging to a subgroup at depth d in the above directed tree has latency $g_d(x) := f(s)/(1+\varepsilon)^d$;
- there are o further self-loops on each resource of type n;
- the out-degree (resp. in-degree) of any resource of subgroups of type $j \in [n]_0$ is j (resp. o), excluding (resp. including) the contribution of self-loops; in particular, each resource of type $j \in [n]$ has exactly one out-going edge directed to a resource of type h for any $h \in [j-1]_0$, and each resource of type $h \in [n-1]_0$ has exactly o in-going edges from o distinct resources of type j, for some $j \in [n]_{h+1}$.

One can show that a load-balancing graph verifying the above properties exists.

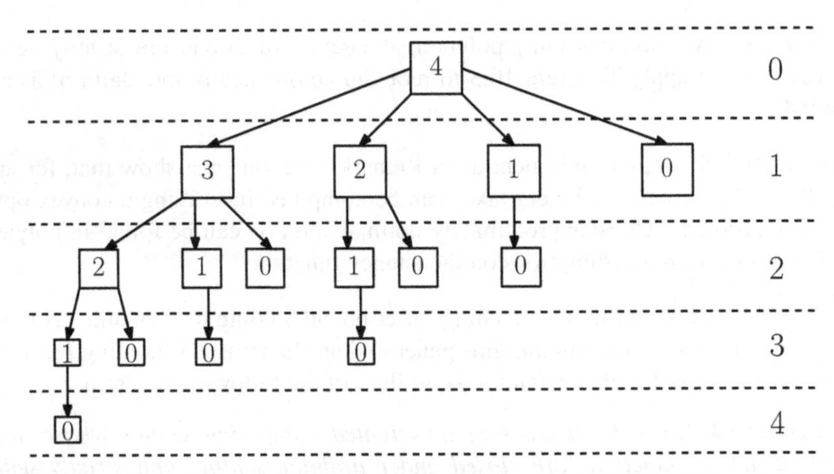

Fig. 10.2 The subgroup structure of the lower bound instance of Theorem 10.4, with $n = 4$ and a generic $o > 0$. In particular, the directed tree depicted in the figure is the subgraph induced by subgroups of nodes/resources in $\mathsf{LB}(4,o)$. Each block represents a subgroup of resources, and the number associated with each block represents its type. The numbers on the right denote the depth of the corresponding subgroups on the left. Each directed edge going from subgroup a to subgroup b means that there are edges from resources of a to resources of b in the underlying load-balancing graph. We observe that the type of each subgroup at depth $d \geq 1$ equivalently represents the congestion in $\boldsymbol{\sigma}$ of the corresponding resources. Instead, the congestion in $\boldsymbol{\sigma}$ of the resources in the unique subgroup at depth $d = 0$ is equal to $n + o$ (because of the self-loop players selecting such resources). Refer to the example represented in Figure 10.3 to see a zoomed part of the load-balancing graph, and in particular, how the edges connect the resources from different subgroups.

Let $\boldsymbol{\sigma}$ and $\boldsymbol{\sigma}^*$ be the strategy profiles in which all players play their first and their second strategy, respectively. By construction, we have that $k_e(\boldsymbol{\sigma}) = j$ for any resource e of type $j \in [n-1]_0$, $k_e(\boldsymbol{\sigma}) = n + o$ for the resources e of type n, and

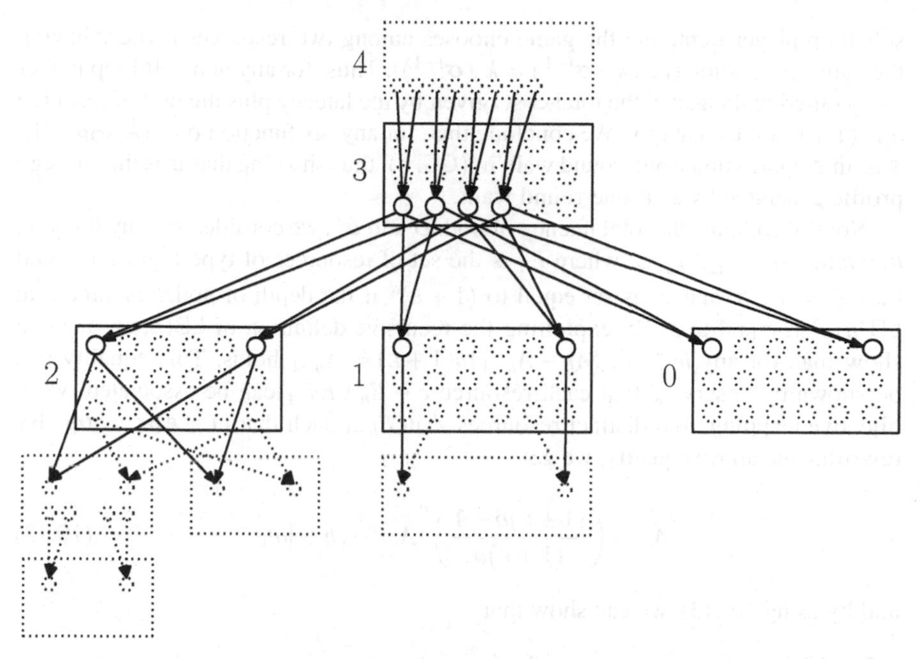

Fig. 10.3 A more detailed description of a portion of the load balancing graph $\mathsf{LB}(n,2)$ (i.e., with a generic n and $o = 2$), made of four nodes/resources from a subgroup of type 3, and their child nodes/resources from subgroups of type $2,1,0$. We observe that, according to the recursive definition of $\mathsf{LB}(n,2)$, each resource of type 3 has exactly one out-going edge directed to a resource of type h for any $h \in \{0,1,2\}$, and each resource of type $h \in \{0,1,2\}$ has exactly 2 in-going edges from 2 distinct resources of type 3.

$k_e(\boldsymbol{\sigma}^*) = o$ for any resource $e \in E$ (regardless of the type). See Figures 10.2 and 10.3 for a clarifying example.

Let $\boldsymbol{\tau}$ be the online process defined as follows: (i) all non-self-loop players enter the game in non-decreasing order with respect to the type of the subgroup which their second strategy/resource belongs to (ties are broken arbitrarily), but (ii) they choose their first strategy; (iii) finally, all the self-loop players select the resources of type n which they are associated with. We observe that $\boldsymbol{\sigma}$ is the strategy profile generated by $\boldsymbol{\tau}$.

In the following, we show that $\boldsymbol{\tau}$ is an ε-one-round-walk, regardless of the considered tax function. By exploiting the strict semi-convexity of the latencies, one can prove that, for any optimal strategy profile, the congestion of each resource $e \in E$ is necessarily equal to $o = k_e(\boldsymbol{\sigma}^*)$. Thus, analogously to Theorem 10.2, we can consider without loss of generality tax functions based on $\boldsymbol{\sigma}^*$.

Fix a tax function $\delta \in \mathsf{TAX}_{\mathsf{OPT}}(\mathcal{G})$. Since the considered tax function is based on the social optimum, we have that δ_e is defined as $\delta_e(k) := \hat{\delta}(k,o,f)/(1+\varepsilon)^d$ for any resource e of a subgroup at depth d, where $\hat{\delta}$ is a function that only depends on the congestion k of the resource e (with respect to the considered strategy profile), the optimal congestion $o = k_e(\boldsymbol{\sigma}^*)$ and the function f. By construction of $\boldsymbol{\tau}$, any non-self-loop player i entering the game chooses among two resources u and v having the same congestion (i.e., $k_u(\boldsymbol{\sigma}^{i-1}) = k_v(\boldsymbol{\sigma}^{i-1})$). Thus, for any non-self loop player (u,v) entering the game, the total cost (given by the latency plus the tax) of resource u is $(1+\varepsilon)$ times that of v. We conclude that, for any tax function $\delta \in \mathsf{TAX}_{\mathsf{OPT}}(\mathcal{G})$, $\boldsymbol{\tau}$ is an ε-approximate one-round walk in (CG,δ), thus showing that $\boldsymbol{\sigma}$ is the strategy profile generated by an ε-one-round walk.

Now, to compute the total latency of both $\boldsymbol{\sigma}$ and $\boldsymbol{\sigma}^*$, we consider, for any $h \in [n]_0$, the value $A_h := \sum_{e \in E_h} \alpha_e$, where E_h is the set of resources of type higher or equal than $j := n-h$, and α_e is set equal to $(1+\varepsilon)^d$ if the depth of node/resource e in $\mathsf{LB}(n,o)$ is equal to d. By exploiting the recursive definition of $\mathsf{LB}(n,o)$, one can show that, for any $h \in [n]$, $(A_h - A_{h-1})o(1+\varepsilon) = A_{h-1}$ holds. This equality can be shown by observing that each resource $e \in E_h \setminus E_{h-1}$ can be associated, via a bijective mapping, to o distinct resources e' in E_{h-1} such that $(1+\varepsilon)\alpha_e = \alpha_{e'}$. By rewriting the above equality, we get

$$A_h = \left(\frac{(1+\varepsilon)o+1}{(1+\varepsilon)o}\right)^h A_0, \quad \forall h \in [n]_0, \tag{10.13}$$

and by using (10.13), we can show that

$$\begin{aligned}\frac{\mathsf{SUM}(\boldsymbol{\sigma})}{A_0} &= (n+o)f(n+o) + \sum_{h=1}^{n}\left(\frac{A_h - A_{h-1}}{A_0}\right)(n-h)f(n-h)\\ &= (n+o)f(n+o) + \sum_{h=1}^{n}\frac{1}{o(1+\varepsilon)}\left(\frac{o(1+\varepsilon)+1}{o(1+\varepsilon)}\right)^{h-1}(n-h)f(n-h)\end{aligned}$$

and

$$\begin{aligned}\frac{\mathsf{SUM}(\boldsymbol{\sigma}^*)}{A_0} &= \sum_{h=0}^{n}\left(\frac{A_h - A_{h-1}}{A_0}\right)of(o)\\ &= of(o) + \sum_{h=1}^{n}\frac{1}{o(1+\varepsilon)}\left(\frac{o(1+\varepsilon)+1}{o(1+\varepsilon)}\right)^{h-1}of(o).\end{aligned}$$

Then, we have

$$\begin{aligned}\lim_{n\to\infty}\frac{\mathsf{SUM}(\boldsymbol{\sigma})}{\mathsf{SUM}(\boldsymbol{\sigma}^*)} &= \lim_{n\to\infty}\frac{\frac{\mathsf{SUM}(\boldsymbol{\sigma})}{\left(\frac{1}{o(1+\varepsilon)}\left(\frac{o(1+\varepsilon)+1}{o(1+\varepsilon)}\right)^{n-1}\right)A_0}}{\frac{\mathsf{SUM}(\boldsymbol{\sigma}^*)}{\left(\frac{1}{o(1+\varepsilon)}\left(\frac{o(1+\varepsilon)+1}{o(1+\varepsilon)}\right)^{n-1}\right)A_0}}\\ &= \lim_{n\to\infty}\frac{\sum_{h=1}^{n}\left(\frac{o(1+\varepsilon)}{o(1+\varepsilon)+1}\right)^{n-h}(n-h)f(n-h)}{\sum_{h=1}^{n}\left(\frac{o(1+\varepsilon)}{o(1+\varepsilon)+1}\right)^{n-h}of(o)}\end{aligned}$$

$$= \lim_{n\to\infty} \frac{\sum_{j=0}^{n-1}\left(\frac{o(1+\varepsilon)}{o(1+\varepsilon)+1}\right)^j jf(j)}{\sum_{j=0}^{n-1}\left(\frac{o(1+\varepsilon)}{o(1+\varepsilon)+1}\right)^j of(o)}$$
$$= \mathcal{E}_f(1+\varepsilon, o)$$
$$> M.$$

We conclude that, for a sufficiently large n, any refundable tax function based on the social optimum is not M-efficient for $\mathsf{LB}(n,o)$, and this shows the claim. □

Remark 10.6. Similarly to Remark 10.3, the lower bounding instance provided in Theorem 10.4 has been suggested by the dual DLP of Lemma 10.3. In particular, if we transform the taxes of DLP in further dual variables, and then we analyze the resulting primal-dual pair as in Remark 10.3, we get a set of tight constraints that helps to design a tight lower bounding instance.

10.2.2 Non-atomic Games

Similarly to Theorem 10.1, we obtain efficient taxes with respect to non-atomic one-round walks. We first provide a preliminary lemma, that can be seen as a non-atomic version of Lemma 10.3 and follow the proof arguments of Theorem 7.1.

Lemma 10.4. *Let $\mathcal{G}$ be a class of non-atomic congestion games, $\mathsf{NCG} \in \mathcal{G}$ be a fixed congestion game, and let $\delta \in \mathsf{TAX}_{\mathsf{OPT}}(\mathcal{G})$ be a refundable tax for NCG defined as $\delta_e(k) := \delta(k,o,f)$, with o and f being the optimal congestion and the latency function of the considered resource e. We have that δ is γ^*-efficient for NCG if $\delta(k,o,f)$ is non-decreasing in k for any fixed o,f, and the following inequality is satisfied for any $k,o \geq 0$ and $f \in \mathcal{C}(\mathcal{G})$:*

$$\gamma^* of(o) \geq kf(k) - \int_0^k f(t)dt$$
$$- \overline{\delta}(k,o,f) + (1+\varepsilon)o\left(f(k) + \frac{\partial}{\partial k}\overline{\delta}(k,o,f)\right), \tag{10.14}$$

where $\overline{\delta}(k,o,f)$ is defined as $\int_0^k \delta(t,o,f)dt$ (that is, $\frac{\partial}{\partial k}\overline{\delta}(k,o,f) = \delta(k,o,f)$).

Given $o > 0$, a class of latency functions $\mathcal{G}$, and $f \in \mathcal{G}$, let:

$$\mathcal{N}_f(1+\varepsilon, o) := \frac{\int_0^\infty e^{-\frac{1}{(1+\varepsilon)o}t} tf(t)dt}{\int_0^\infty e^{-\frac{1}{(1+\varepsilon)o}t} of(o)dt} = \frac{\int_0^\infty e^{-\frac{1}{(1+\varepsilon)o}t} tf(t)dt}{(1+\varepsilon)o^2 f(o)}, \tag{10.15}$$

$$\mathcal{N}_{\mathcal{G}}(1+\varepsilon) := \sup_{f\in\mathcal{C}(\mathcal{G}), o>0} \mathcal{N}_f(1+\varepsilon, o). \tag{10.16}$$

By using the same proof arguments of Theorem 10.3, we can resort to Lemma 10.4 to get the following theorem.

Theorem 10.5. *Let $\mathcal{G}$ be a class of non-atomic games whose non-constant latency functions are strictly semi-convex. Given* $\mathsf{NCG} \in \mathcal{G}$*, there exists a* $\mathcal{N}_{\mathcal{G}}(1+\varepsilon)$*-efficient refundable tax function* $\delta \in \mathsf{TAX}_{\mathsf{OPT}}(\mathcal{G})$ *for* NCG*, with respect to* ε*-approximate non-atomic one-round walks and to the total latency.*

Proof (Sketch). For any fixed non-constant latency $f \in \mathcal{C}(\mathcal{G})$ and $o > 0$, let $\delta(k,o,f)$ be the tax function defined as follows: (i) $\delta(k,o,f) := 0$ for $k := 0$; (ii) (10.9) holds as equality for the considered o, f and γ^* equal to $\gamma^*_{o,f} := \mathcal{N}_f(1+\varepsilon,o)$. We observe that $\delta(k,o,f)$ can be computed by solving the following Cauchy problem in a real function y defined as $y(k) := \int_0^t \delta(t,o,f)dt$ (that is, $\delta(k,o,f) = \frac{\partial}{\partial k}y(k)$):

$$\begin{cases} \mathcal{N}_f(1+\varepsilon,o)of(o) = kf(k) - \int_0^k f(t)dt - y(t) + (1+\varepsilon)o\left(f(k) + \frac{\partial}{\partial k}y(k)\right) \\ y(0) = 0. \end{cases}$$

If f is constant or $o = 0$, we set $\delta(k,o,f) := 0$ for any k.

By proceeding as in the proof of Theorem 10.2, one can show that δ is non-negative and non-decreasing. Furthermore, one can show that (10.14) is satisfied by δ. We conclude that the hypotheses of Lemma 10.4 are satisfied, and then δ is $\mathcal{N}_{\mathcal{G}}(1+\varepsilon)$-efficient refundable. □

Remark 10.7. As (non-constant) polynomial latency functions are strictly semi-convex, we can apply Theorem 10.5 to reobtain automatically the claim of Theorem 9.7.

Remark 10.8. For any given non-atomic congestion game with semi-convex latency functions, under mild assumptions, we can find in polynomial time a $(1+\xi)$-approximate optimal solution for any fixed $\xi > 0$. Thus, the same arguments of Theorem 10.5 can be used to derive $(\mathcal{N}_{\mathcal{G}}(1+\varepsilon)+\xi)$-efficient taxes based on such approximate solutions.

In the following theorem, we show that the efficiency shown in Theorem 10.5 cannot be improved, when considering taxes based on optimal strategy profiles.

Theorem 10.6. *Let $\mathcal{G}$ be a class of unweighted congestion games whose latencies are closed under ordinate scaling and strictly semi-convex. For any fixed $M < \mathcal{N}_{\mathcal{G}}(1+\varepsilon)$, there exists* $\mathsf{NCG} \in \mathcal{G}$ *such that, for any* $\delta \in \mathsf{TAX}_{\mathsf{OPT}}(\mathcal{G})$*, δ is not an M-efficient refundable tax for* CG *with respect to ε-approximate one-round walks and to the total latency. Furthermore, the considered lower bound, holds even for load balancing games, and for load balancing games with identical resources if $\varepsilon = 0$.*

Proof (Sketch). A tight lower bounding instance can be obtained by exploiting the proof of Theorem 10.4. Let $f \in \mathcal{C}(\mathcal{G})$ and $o > 0$ be such that $\mathcal{N}_f(1+\varepsilon,o) > M$, and let $k,t > o$ be sufficiently large real numbers. Let $\mathsf{LB}_f(n',o')$ be the lower bounding instance provided in Theorem 10.4, with $n' := \lceil kt \rceil$, $o' := \lceil ot \rceil$, and with players weights equal to $1/t$. Similarly as in Theorem 7.3, if t is sufficiently large, the considered unweighted lower bounding instance can be transformed in a non-atomic load balancing instance whose competitive ratio, even with the use of taxes, is arbitrarily close to $\mathcal{N}_f(1+\varepsilon,o)$ as t and k increase.

10.3 Concluding Remarks and Related Work

This chapter is based on some preliminary results included in the second author's PhD thesis. In previous works, the use of dynamic taxes to improve the performance in congestion games with general latency functions has been mainly investigated for non-atomic games and pure Nash equilibria, and it has been shown that either universal dynamic taxes (i.e., dynamically dependent on the local congestion of each resource), or static taxes depending on the social optimum, guarantee that the social optimum becomes a pure Nash equilibrium (see Beckmann et al. [17], Pigou [150] for further details). Analogously to non-atomic games and pure Nash equilibria, our results characterize the best taxation mechanisms based on optimal strategy profiles for atomic games and several solution concepts.

Similar results as those obtained for exact Nash equilibria (i.e., the results of Section 10.1 with $\varepsilon = 0$) have been independently achieved by Paccagnan and Gairing [137]. Regarding exact Nash equilibria, their work constitutes a more complete and detailed version of the results discussed in this chapter. In particular, Paccagnan and Gairing [137] show how to compute $\mathcal{B}_f(1,o)$-efficient refundable taxes based on the social optimum, and how to compute them in polynomial time at the expense of an arbitrarily small efficiency loss. Furthermore, they focus on the optimization problem of finding social optima in congestion games, and show that the approximation guaranteed by the above taxes is the best possible, unless $P = NP$.

The works of Paccagnan et al. [140] and Ravindran Vijayalakshmi and Skopalik [152] focus on the framework of dynamic universal taxes. In particular, they independently provide efficient dynamic universal taxes for unweighted games with general latency functions, and show their optimality via tight lower bounds.

Chapter 11
Stackelberg Strategies for Affine Unweighted Congestion Games

In Stackelberg games, a central authority, called the *leader*, is granted the power of dictating the strategies of a subset of players, called the *followers*. The leader's purpose is to determine a good *Stackelberg strategy*, which is an algorithm that carefully chooses a subset of players (called *coordinated players*) and their assigned strategies, so as to mitigate as much as possible the effects caused by the selfish behavior of the *uncoordinated players*, that is, to lower as much as possible the price of anarchy of the resulting game. The power of the mechanism is limited by a parameter $\alpha \in (0,1)$, that is the maximum fraction of players the leader can control. Thus, the higher the α, the lower the resulting price of anarchy.

In this chapter, we consider the application of three Stackelberg strategies, namely Largest Latency First, Cover and Scale, to congestion games with affine latency functions, and we show upper and lower bounds on their worst-case price of anarchy. In particular, by resorting to the primal-dual method, we obtain the following results: for Largest Latency First, we show that the price of anarchy is exactly $(20-11\alpha)/8$ for $\alpha \in [0,4/7]$ and $(4-3\alpha+\sqrt{4\alpha-3\alpha^2})/2$ for $\alpha \in [4/7,1]$; for λ-Cover, we show that the price of anarchy is exactly $\frac{4\lambda-1}{3\lambda-1}$ for affine latency functions and exactly $1+(4\lambda+1)/(4\lambda(2\lambda+1))$ for linear ones; finally, for Scale, we give an upper bound of $1+((1-\alpha)(2h+1))/((1-\alpha)h^2+\alpha h+1)$, where h is the unique integer such that $\alpha \in [(2h^2-3)/(2(h^2-1)),(2h^2+4h-1)/(2h(h+2)))$.

11.1 Stackelberg Strategies: Preliminaries

Assume that there is a central authority, which in this setting is usually referred to as the *leader*, who has the power of dictating the strategies of a subset of players. The leader's purpose is to determine a good *Stackelberg strategy*, which is an algorithm that carefully chooses the subset of players (called *coordinated players*) and their assigned strategies, so as to mitigate as much as possible the effects caused by the selfish behavior of the *uncoordinated players*, that is, to lower as much as possible the price of anarchy of the resulting game.

V. Bilò, C. Vinci, *Coping with Selfishness in Congestion Games*, Monographs in Theoretical Computer Science. An EATCS Series, https://doi.org/10.1007/978-3-031-30261-9_11

Given a congestion game CG and a Stackelberg strategy A, let P_A be the set of coordinated players chosen by A and S_i be the prescribed strategy assigned by A to each player $i \in P_A$. The congestion game CG_A, obtained from CG by coordinating the choices of the players in P_A, is the same as CG with the only difference that, for each player $i \in P_A$, CG_A has $\Sigma_i = \{S_i\}$, that is, each coordinated player becomes a selfish player who has no alternatives except for her prescribed strategy.

A Stackelberg strategy is *optimal-restricted* if the coordinated players are assigned the strategy they adopt in a social optimum $\boldsymbol{\sigma}^*$. For each optimal-restricted Stackelberg strategy A, since the optimal value of CG_A is equal to that of CG, the price of anarchy of CG_A becomes $\mathsf{PoA}(\mathsf{CG}_A) = \max_{\boldsymbol{\sigma} \in \mathsf{NE}(\mathsf{CG}_A)} \frac{\mathsf{SUM}(\boldsymbol{\sigma})}{\mathsf{SUM}(\boldsymbol{\sigma}^*)}$.

In this chapter, we focus on three optimal-restricted Stackelberg strategies. The first two ones are deterministic, while the latter is randomized, hence, its price of anarchy will be evaluated in expectation. Moreover, in the first and third strategy the number of coordinated players is assumed to be equal to αn for some $\alpha \in [0,1]$ with $\alpha n \in \mathbb{N}$, i.e., the leader controls a fraction α of the players in the game. For a fixed congestion game CG and social optimum $\boldsymbol{\sigma}^*$, they are defined as follows:

- $\mathsf{LLF}(\alpha)$: for each $\alpha \in [0,1]$ such that $\alpha n \in \mathbb{N}$, the set of coordinated players is chosen equal to the set of the αn players with the highest cost in $\boldsymbol{\sigma}^*$, breaking ties arbitrarily. LLF stands for Largest Latency First.
- λ-Cover: for each $\lambda \in \mathbb{N} \setminus \{0\}$, the set of coordinated players P is chosen so as to guarantee that $|\{i \in P : j \in \sigma_i^*\}| \geq \min\{\lambda, n_j(\boldsymbol{\sigma}^*)\}$ for each $j \in E$. Note that one such a set P might not exist in CG, so that this strategy is not always applicable.
- $\mathsf{Scale}(\alpha)$: for each $\alpha \in [0,1]$ such that $\alpha n \in \mathbb{N}$, the set of coordinated players is randomly chosen with uniform probability among the set of all subsets of N having cardinality αn.

For a given optimal-restricted Stackelberg strategy A, our aim is to focus on the characterization of the price of anarchy of game CG_A, when CG is an affine congestion game. Let $\mathcal{G}$ be any subclass of the class of congestion games. We define the price of anarchy of Stackelberg strategy A in the class of games $\mathcal{G}$ as $\mathsf{PoA}_{\mathcal{G}}(A) = \sup_{\mathsf{CG} \in \mathcal{G}} \mathsf{PoA}(\mathsf{CG}_A)$. Now let $\mathcal{ACG}$ and $\mathcal{LCG}$ be the class of affine congestion games and that of linear congestion games, respectively. Let $\ell_j(x) := a_j x + b_j$ (resp. $\ell(x) := a_j x$) denote the latency of resource $j \in E$ of an affine (resp. linear) congestion game. By exploiting a result in [22], it follows that for each affine congestion game CG, there exists a linear congestion game CG′, defined on a different set of resources, with $\mathsf{PoA}(\mathsf{CG}'_A) = \mathsf{PoA}(\mathsf{CG}_A)$. Hence, it is possible to conclude that, for any Stackelberg strategy $A \in \{\mathsf{LLF}(\alpha), \mathsf{Scale}(\alpha)\}$,

$$\mathsf{PoA}_{\mathcal{ACG}}(A) = \mathsf{PoA}_{\mathcal{LCG}}(A). \tag{11.1}$$

As a consequence of (11.1), we will be allowed to restrict our attention only to the class of linear congestion games when bounding the price of anarchy of these two

Stackelberg strategies in the class of affine congestion games.[1] Moreover, we will use the simplified notation $\mathsf{PoA}(A)$ to denote the value $\mathsf{PoA}_{\mathcal{ACG}}(A) = \mathsf{PoA}_{\mathcal{LCG}}(A)$ when considering $A \in \{\mathsf{LLF}(\alpha), \mathsf{Scale}(\alpha)\}$. However, we cannot exploit the result in [22] to prove (11.1) in the case in which $A = \lambda$-Cover, thus, in this case, we consider the cases of affine and linear latency functions separately.

11.2 Price of Anarchy of Largest Latency First

In this section, we provide a tight bound on the price of anarchy of Largest Latency First. For the characterization of the upper bound, we use the primal-dual method. To this aim, fix a linear congestion game CG, a social optimum $\boldsymbol{\sigma}^*$ and a pure Nash equilibrium $\boldsymbol{\sigma}$ induced by $\mathsf{LLF}(\alpha)$. Let P be the set of αn coordinated players chosen by $\mathsf{LLF}(\alpha)$. For the sake of conciseness, for each $j \in E$, we set $x_j := n_j(\boldsymbol{\sigma})$, $o_j := n_j(\boldsymbol{\sigma}^*)$ and $s_j := |\{i \in P : j \in \sigma_i^*\}|$. Clearly, we have $s_j \leq \min\{x_j, o_j\}$ for each $j \in E$.

We obtain the following primal linear program $\mathsf{LP} := \mathsf{LP}(\boldsymbol{\sigma}, \boldsymbol{\sigma}^*)$ defined over the latency functions of the resources in E:

$$
\begin{aligned}
\mathsf{LP}: \quad & \max \sum_{j \in E} a_j x_j^2 \\
s.t. \quad & \sum_{j \in \sigma_i} a_j x_j - \sum_{j \in \sigma_i^*} a_j (x_j + 1) \leq 0, \quad \forall i \in N \setminus P \\
& \sum_{j \in \sigma_i^*} a_j o_j - \sum_{j \in \sigma_k^*} a_j o_j \leq 0, \quad \forall i \in N \setminus P, k \in P \\
& \sum_{j \in E} a_j o_j^2 = 1 \\
& a_j \geq 0, \quad \forall j \in E,
\end{aligned}
$$

where the first family of constraints is satisfied since each uncoordinated player cannot lower her cost by switching to the strategy she adopts in $\boldsymbol{\sigma}^*$ and the second one is satisfied because $\mathsf{LLF}(\alpha)$ selects the set of $|P| = \alpha n$ players with the highest cost in $\boldsymbol{\sigma}^*$.

The dual program $\mathsf{DLP} := \mathsf{DLP}(\boldsymbol{\sigma}, \boldsymbol{\sigma}^*)$, obtained by associating the variables $(y_i)_{i \in N \setminus P}$ with the first family of constraints, the variables $(z_{ik})_{i \in N \setminus P, k \in P}$ with the second family of constraints and the variable γ with the last constraint, is the following:

$$
\mathsf{DLP}: \quad \min \gamma
$$

[1] This equivalence does not apply to λ-Cover, as, since the transformation given in [22] changes the set of resources, a choice inducing a λ-Cover for game CG might not guarantee a λ-Cover for game CG'.

$$
\begin{aligned}
s.t. \sum_{i \in N\setminus P: j \in \sigma_i} y_i x_j - \sum_{i \in N\setminus P: j \in \sigma_i^*} y_i(x_j+1) + \\
+ \sum_{i \in N\setminus P, k \in P: j \in \sigma_i^*} z_{ik} o_j - \sum_{i \in N\setminus P, k \in P: j \in \sigma_k^*} z_{ik} o_j + \gamma o_j^2 \geq x_j^2, \quad \forall j \in E \\
y_i \geq 0, \quad \forall i \in N \setminus P \\
z_{ik} \geq 0, \quad \forall i \in N \setminus P, k \in P.
\end{aligned}
$$

Any feasible solution for DLP which is independent of the choices of $\boldsymbol{\sigma}$ and $\boldsymbol{\sigma}^*$, i.e., satisfying the dual constraint for each possible pair of integers $(x_j, o_j) \in \{N \cup \{0\}\}^2$, will provide an upper bound on $\mathsf{PoA}(\mathsf{CG}_{\mathsf{LLF}(\alpha)})$ which, by the generality of CG, gives an upper bound on $\mathsf{PoA}(\mathsf{LLF}(\alpha))$. We get the following result.

Theorem 11.1. *For each $\alpha \in [0,1]$,*

$$
\mathsf{PoA}(\mathsf{LLF}(\alpha)) \leq \begin{cases} \frac{20-11\alpha}{8} & \text{for } \alpha \in [0, 4/7], \\ \frac{4-3\alpha+\sqrt{4\alpha-3\alpha^2}}{2} & \text{for } \alpha \in [4/7, 1]. \end{cases}
$$

Proof. Set $y_i = \frac{3}{2}$ for each $i \in N \setminus P$, $z_{ik} = \frac{11}{8n}$ for each $i \in N \setminus P$ and $k \in P$, and $\gamma = \frac{20-11\alpha}{8}$. With these values, the generical dual constraint becomes

$$
\begin{aligned}
\frac{3}{2}\left(\sum_{i \in N\setminus P: j \in \sigma_i} x_j - \sum_{i \in N\setminus P: j \in \sigma_i^*} x_j + 1\right) + \\
+ \frac{11}{8n}\left(\sum_{i \in N\setminus P, k \in P: j \in \sigma_i^*} o_j - \sum_{i \in N\setminus P, k \in P: j \in \sigma_k^*} o_j\right) + \frac{20-11\alpha}{8} o_j^2 \geq x_j^2,
\end{aligned}
$$

which, using

$$
|\{i \in N \setminus P : j \in \sigma_i\}| = x_j - s_j, \tag{11.2}
$$

$$
|\{i \in N \setminus P : j \in \sigma_i^*\}| = o_j - s_j, \tag{11.3}
$$

$$
|\{i \in N \setminus P, k \in P : j \in \sigma_i^*\}| = \alpha n(o_j - s_j), \tag{11.4}
$$

$$
|\{i \in N \setminus P, k \in P : j \in \sigma_k^*\}| = (1-\alpha) n s_j, \tag{11.5}
$$

becomes

$$
f(x_j, o_j, s_j) := 4x_j^2 - 12x_j o_j - 12o_j + 12s_j - 11o_j s_j + 20o_j^2 \geq 0.
$$

We show that $f(x_j, o_j, s_j) \geq 0$ for any triple of non-negative integers (x_j, o_j, s_j) such that $s_j \leq \min\{x_j, o_j\}$. Let us first compute the derivative $\frac{\delta f}{\delta x_j}(x_j, o_j, s_j) = 8x_j - 12o_j$. This is a linear function on x_j which is negative for $x_j = 0$, hence $f(x_j, o_j, s_j)$ is minimized for $x_j = \frac{3o_j}{2}$. By substituting, we get that $f(x_j, o_j, s_j) \geq 0$ if

$$g(o_j,s_j) := f\left(\frac{3o_j}{2},o_j,s_j\right) = 11o_j^2 - 11o_js_j - 12o_j + 12s_j \geq 0.$$

Again, let us compute the derivative $\frac{\delta g}{\delta s_j}(o_j,s_j) = 12 - 11o_j$. For $o_j \geq 2$, the derivative is always negative which implies that $g(o_j,s_j)$ is minimized for $s_j = o_j$. By substituting, we get $g(o_j,o_j) \geq 0$ as desired. For the leftover cases of $o_j \leq 1$, we have $f(x_j,0,0) = 4x_j^2 \geq 0$ and $f(x_j,1,s_j) = 4x_j^2 - 12x_j + s_j + 8 \geq 0$ as desired.

For the second bound, set $\theta = \sqrt{4\alpha - 3\alpha^2}$, $y_i = 1 + \frac{\alpha}{\theta}$ for each $i \in N \setminus P$, $z_{ik} = \frac{1}{2n}\left(3 + \frac{3\alpha-2}{\theta}\right)$ for each $i \in N \setminus P$ and $k \in P$, and $\gamma = \frac{4-3\alpha+\theta}{2}$. With these values, by using Equations (11.2)-(11.5), the generical dual constraint becomes

$$\begin{aligned}
&f(x_j,o_j,s_j)\\
&:= 2\alpha x_j^2 - 2(\alpha+\theta)(x_jo_j + o_j - s_j) - (3\theta + 3\alpha - 2)o_js_j + 2(\alpha + 2\theta)o_j^2\\
&\geq 0.
\end{aligned}$$

We show that $f(x_j,o_j,s_j) \geq 0$ for any triple of non-negative integers (x_j,o_j,s_j) such that $s_j \leq \min\{x_j,o_j\}$. Let us first compute the derivative $\frac{\delta f}{\delta x_j}(x_j,o_j,s_j) = 4\alpha x_j - 2(\alpha+\theta)o_j$. This is a linear function on x_j which is negative for $x_j = 0$, hence $f(x_j,o_j,s_j)$ is minimized for $x_j = \frac{(\alpha+\theta)o_j}{2\alpha}$. By substituting, we get that $f(x_j,o_j,s_j) \geq 0$ if

$$\begin{aligned}
&g(o_j,s_j)\\
&:= f\left(\frac{(\alpha+\theta)o_j}{2\alpha},o_j,s_j\right)\\
&= 2(3\alpha+3\theta-2)o_j^2 - 2((3\alpha+3\theta-2)s_j + 2(\alpha+\theta))o_j + 4(\alpha+\theta)s_j\\
&\geq 0.
\end{aligned}$$

Again, let us compute the derivative $\frac{\delta g}{\delta s_j}(o_j,s_j) = 4(\alpha+\theta) - 2(3\alpha+3\theta-2)o_j$. Since $\alpha+\theta-1 \geq 0$ for each $\alpha \in [4/7,1]$, this function is non-increasing for $o_j \geq 2$, so that $g(o_j,s_j)$ is minimized for $s_j = o_j$ when $o_j \geq 2$. By substituting, we get $g(o_j,o_j) \geq 0$ as desired. For the leftover cases of $o_j \leq 1$, we have $f(x_j,0,0) = 2\alpha x_j^2 \geq 0$ and $f(x_j,1,s_j) = 2\alpha x_j^2 - 2(\alpha+\theta)x_j + (2-\alpha-\theta)s_j + 2\theta$. Since $2 - \alpha - \theta \leq 0$ for each $\alpha \in [4/7,1]$, we have that this last function is minimized for $s_j = 0$. Hence, we need to show that $f(x_j,1,0) = 2\alpha x_j^2 - 2(\alpha+\theta)x_j + 2\theta \geq 0$. By solving for x_j, we get $f(x_j,1,0) \geq 0$ when $x_j \leq 1$ or $x_j \geq \theta/\alpha$. Since, $\theta/\alpha \leq 2$ for each $\alpha \in [4/7,1]$, $f(x_j,1,0) \geq 0$ for each non-negative integer x_j and the proof is complete. □

We now prove matching lower bounds.

Theorem 11.2. *For each* $\alpha \in [0,4/7]$, $\mathsf{PoA}(\mathsf{LLF}(\alpha)) \geq \frac{20-11\alpha}{8}$, *while for each* $\alpha \in [4/7,1]$, $\mathsf{PoA}(\mathsf{LLF}(\alpha)) \geq \frac{4-3\alpha+\sqrt{4\alpha-3\alpha^2}}{2}$.

Proof. For a fixed $\alpha \in [0,4/7]$, let n be an integer such that $\alpha n \in \mathbb{N}$, $\alpha n \geq 2$ and $(1-\alpha)n \geq 3$. Consider the linear congestion game CG defined as follows.

The set of players is $N = N_1 \cup N_2$, where $|N_1| = (1-\alpha)n$ and $|N_2| = \alpha n$.

The set of resources is $E = U \cup V \cup W$, where $U = \{u_j : j \in [(1-\alpha)n]\}$, $V = \{v_j : j \in [(1-\alpha)n]\}$ and $W = \{w_{j,k} : j \in [(1-\alpha)n], k \in [\alpha n]\}$. Each resource in U has latency function $\ell_U(x) = \frac{(4-\alpha)n}{2}x$, each resource in V has latency function $\ell_V(x) = \frac{(4-7\alpha)n}{2}x$ and each resource in W has latency function $\ell_W(x) = x$. Note that $\alpha \in [0,4/7]$ guarantees $\frac{(4-7\alpha)n}{2} \geq 0$.

Each player $i \in N_1$ has two strategies: the *first strategy* $\{u_{i+1}, v_{i+1}, v_{i+2}\} \cup \bigcup_{k\in[\alpha n]}\{w_{i,k}\}$ and the *second strategy* $\{u_i, v_i\}$, where the sums over the indices have to be interpreted circularly so that, for instance, $u_{(1-\alpha)n+1} = u_1$. Each player $i \in N_2$ has only one available strategy given by $\bigcup_{k\in[(1-\alpha)n]}\{w_{k,i}, w_{k+1,i}\}$, where again the sums over the indices have to be interpreted circularly so that, for instance, $w_{(1-\alpha)n+1,i} = w_{1,i}$. Note that, by $\alpha n \geq 2$, the number of resources in W belonging to the strategy of each player in N_2 is exactly $2(1-\alpha)n$; moreover, by $(1-\alpha)n \geq 3$, the first and second strategy of each player in N_1 are disjoint.

First of all, it is not difficult to see that the strategy profile $\boldsymbol{\sigma}^*$ in which each player in N_1 adopts her second strategy is a social optimum for CG. Now note that, for each player $i \in N_1$, $c_i(\boldsymbol{\sigma}^*) = \frac{(4-\alpha)n}{2} + \frac{(4-7\alpha)n}{2} = 4(1-\alpha)n$, whereas, for each player $i \in N_2$, $c_i(\boldsymbol{\sigma}^*) = 4(1-\alpha)n$ since each player in N_2 uses $2(1-\alpha)n$ resources in W each having congestion equal to 2. Hence, it follows that $\mathsf{LLF}(\alpha)$ may choose N_2 as the set of coordinated players. We now show that, under this hypothesis, the strategy profile $\boldsymbol{\sigma}$ in which each player $i \in N_1$ adopts her first strategy is a pure Nash equilibrium for $\mathsf{CG}_{\mathsf{LLF}(\alpha)}$.

For each player $i \in N_1$, $c_i(\boldsymbol{\sigma}) = \frac{(4-\alpha)n}{2} + 4\frac{(4-7\alpha)n}{2} + 3\alpha n = \frac{(20-23\alpha)n}{2}$ since each player $i \in N_1$ uses 1 resource in U having congestion 1, 2 resources in V both having congestion 2 and αn resources in W each having congestion 3. Let $\boldsymbol{\sigma}'$ be the strategy profile obtained from $\boldsymbol{\sigma}$ when player i deviates to her second strategy. In this case, $c_i(\boldsymbol{\sigma}') = 2\frac{(4-\alpha)n}{2} + 3\frac{(4-7\alpha)n}{2} = \frac{(20-23\alpha)n}{2}$ since player i uses 1 resource in U having congestion 2 and 1 resource in V having congestion 3. Hence, it follows that $\boldsymbol{\sigma}$ is a pure Nash equilibrium for $\mathsf{CG}_{\mathsf{LLF}(\alpha)}$.

We are now left to compute the ratio $\frac{\mathsf{SUM}(\boldsymbol{\sigma})}{\mathsf{SUM}(\boldsymbol{\sigma}^*)}$. To this aim, note that $n_j(\boldsymbol{\sigma}) = n_j(\boldsymbol{\sigma}^*) = 1$ for each resource $j \in U$, $n_j(\boldsymbol{\sigma}) = 2$ and $n_j(\boldsymbol{\sigma}^*) = 1$ for each resource $j \in V$, while, for each resource $j \in W$, $n_j(\boldsymbol{\sigma}) = 3$ and $n_j(\boldsymbol{\sigma}^*) = 2$. We get

$$\frac{\mathsf{SUM}(\boldsymbol{\sigma})}{\mathsf{SUM}(\boldsymbol{\sigma}^*)} = \frac{(1-\alpha)n\left(\frac{(4-\alpha)n}{2} + 4\frac{(4-7\alpha)n}{2}\right) + 9\alpha(1-\alpha)n^2}{(1-\alpha)n\left(\frac{(4-\alpha)n}{2} + \frac{(4-7\alpha)n}{2}\right) + 4\alpha(1-\alpha)n^2}$$
$$= \frac{20-11\alpha}{8}$$

and the claim follows because $\mathsf{PoA}(\mathsf{LLF}(\alpha)) \geq \mathsf{PoA}(\mathsf{CG}_{\mathsf{LLF}(\alpha)})$.

For a fixed $\alpha \in [4/7, 1]$, set $\theta = \sqrt{4\alpha - 3\alpha^2}$ and let n be an integer such that $\frac{\alpha n}{2} \in \mathbb{N}$, $\frac{(\theta-\alpha)n}{4} \in \mathbb{N}$ and $(1-\alpha)n \geq 2$. Note that, for each $\alpha \in [4/7, 1]$, $\theta - \alpha \geq 0$ so that n is well defined. Consider the linear congestion game CG defined as follows.

The set of players is $N = N_1 \cup N_2$, where $|N_1| = (1-\alpha)n$ and $|N_2| = \alpha n$.

The set of resources is $E = U \cup W$, where $U = \{u_j : j \in [(1-\alpha)n]\}$ and $W = \{w_{j,k} : j \in [(1-\alpha)n], k \in [\alpha n]\}$. Each resource in U has latency function $\ell_U(x) = \frac{\alpha^2(1-\alpha)n^3}{4} x$ and each resource in W has latency function $\ell_W(x) = x$.

Each player $i \in N_1$ has two available strategies: the *first strategy* $\{u_{i+1}\} \cup \bigcup_{k\in[\alpha n], j\in\left[\frac{(\theta-\alpha)n}{4}\right]} \{w_{i+j-1,k}\}$ and the *second strategy* $\{u_i\}$, where again the sums over the indexes have to be interpreted circularly. Note that, for each $\alpha \in [4/7, 1]$, $\frac{(\theta-\alpha)n}{4} \leq (1-\alpha)n$ so that the number of resources in W belonging to the first strategy of each player in N_1 is exactly $\frac{\alpha(\theta-\alpha)n^2}{4}$; moreover, by $(1-\alpha)n \geq 2$, the first and second strategy of each player are disjoint. Each player $i \in N_2$ has only one available strategy given by $\bigcup_{k\in[(1-\alpha)n], j\in\left[\frac{\alpha n}{2}\right]} \{w_{k,i+j-1}\}$, where again the sums over the indexes have to be interpreted circularly.

First of all, it is not difficult to see that the strategy profile $\boldsymbol{\sigma}^*$ in which each player in N_1 adopts her second strategy is a social optimum for CG. Now note that, for each player $i \in N_1$, $c_i(\boldsymbol{\sigma}^*) = \frac{\alpha^2(1-\alpha)n^3}{4}$, whereas, for each player $i \in N_2$, $c_i(\boldsymbol{\sigma}^*) = \frac{\alpha^2(1-\alpha)n^3}{4}$ since each player in N_2 uses $\frac{\alpha n}{2}(1-\alpha)n$ resources in W each having congestion equal to $\frac{\alpha n}{2}$. Hence, it follows that $\mathsf{LLF}(\alpha)$ may choose N_2 as the set of coordinated players. We now show that, under this hypothesis, the strategy profile $\boldsymbol{\sigma}$ in which each player $i \in N_1$ adopts her first strategy is a pure Nash equilibrium for $\mathsf{CG}_{\mathsf{LLF}(\alpha)}$.

For each player $i \in N_1$, $c_i(\boldsymbol{\sigma}) = \frac{\alpha^2(1-\alpha)n^3}{4} + \frac{\alpha(\theta-\alpha)n^2}{4}\frac{(\theta+\alpha)n}{4} = \frac{\alpha^2(1-\alpha)n^3}{2}$ since each player $i \in N_1$ uses 1 resource in U having congestion 1 and $\frac{\alpha(\theta-\alpha)n^2}{4}$ resources in W each having congestion $\frac{(\theta+\alpha)n}{4}$. Let $\boldsymbol{\sigma}'$ be the strategy profile obtained from $\boldsymbol{\sigma}$ when player i deviates to her second strategy. In this case, $c_i(\boldsymbol{\sigma}') = 2\frac{\alpha^2(1-\alpha)n^3}{4} = \frac{\alpha^2(1-\alpha)n^3}{2}$, since player i uses 2 resources in U both having congestion 2. Hence, it follows that $\boldsymbol{\sigma}$ is a pure Nash equilibrium for $\mathsf{CG}_{\mathsf{LLF}(\alpha)}$.

We are now left to compute the ratio $\frac{\mathsf{SUM}(\boldsymbol{\sigma})}{\mathsf{SUM}(\boldsymbol{\sigma}^*)}$. To this aim, note that $n_j(\boldsymbol{\sigma}) = n_j(\boldsymbol{\sigma}^*) = 1$ for each resource $j \in U$, while, for each resource $j \in W$, $n_j(\boldsymbol{\sigma}) = \frac{(\theta+\alpha)n}{4}$ and $n_j(\boldsymbol{\sigma}^*) = \frac{\alpha n}{2}$. We get

$$\frac{\mathsf{SUM}(\boldsymbol{\sigma})}{\mathsf{SUM}(\boldsymbol{\sigma}^*)} = \frac{(1-\alpha)n\frac{\alpha^2(1-\alpha)n^3}{4} + \alpha(1-\alpha)n^2\frac{(\alpha+\theta)^2n^2}{16}}{(1-\alpha)n\frac{\alpha^2(1-\alpha)n^3}{4} + \alpha(1-\alpha)n^2\frac{\alpha^2n^2}{4}}$$

$$= \frac{4 - 3\alpha + \sqrt{4\alpha - 3\alpha^2}}{2}$$

and the claim follows because $\mathsf{PoA}(\mathsf{LLF}(\alpha)) \geq \mathsf{PoA}(\mathsf{CG}_{\mathsf{LLF}(\alpha)})$. □

11.3 Price of Anarchy of λ-Cover

In this section, we characterize the performance of λ-Cover by showing that, for each $\lambda \in \mathbb{N} \setminus \{0\}$, $\mathsf{PoA}_{\mathcal{A}\mathcal{C}\mathcal{G}}(\lambda\text{-Cover}) = \frac{4\lambda-1}{3\lambda-1}$ and $\mathsf{PoA}_{\mathcal{L}\mathcal{C}\mathcal{G}}(\lambda\text{-Cover}) = 1 + \frac{4\lambda+1}{4\lambda(2\lambda+1)}$. In order to show the upper bound for linear latency functions, we make use again of the primal-dual method. To this aim, fix a linear congestion game CG, a social optimum $\boldsymbol{\sigma}^*$ and a pure Nash equilibrium $\boldsymbol{\sigma}$ induced by λ-Cover applied on a particular choice of the set of coordinated players P. Again, for each $j \in E$, we set $x_j := n_j(\boldsymbol{\sigma})$, $o_j := n_j(\boldsymbol{\sigma}^*)$ and $s_j := |\{i \in P : j \in \sigma_i^*\}|$. In this case, by the definition of λ-Cover, for each $j \in E$, we have

$$s_j \geq \min\{\lambda, o_j\}. \tag{11.6}$$

We obtain the following primal linear program $\mathsf{LP} := \mathsf{LP}(\boldsymbol{\sigma}, \boldsymbol{\sigma}^*)$:

$$\begin{aligned}
\mathsf{LP}: \quad & \max \sum_{j\in E} a_j x_j^2 \\
& s.t. \quad \sum_{j\in\sigma_i} a_j x_j - \sum_{j\in\sigma_i^*} a_j(x_j+1) \leq 0, \quad \forall i \in N \setminus P \\
& \qquad \sum_{j\in E} a_j o_j^2 = 1 \\
& \qquad a_j \geq 0, \quad \forall j \in E.
\end{aligned}$$

Observe that the proposed program is a subprogram of the one defined for LLF and does not depend on λ. This must not be surprising, since the parameter λ does not affect the players' costs directly, but only gives us an additional property on the resources congestion which can be suitably exploited in the analysis of the dual constraints, see Equation (11.6). The dual program $\mathsf{DLP} := \mathsf{DLP}(\boldsymbol{\sigma}, \boldsymbol{\sigma}^*)$, obtained by associating the variables $(y_i)_{i\in N\setminus P}$ with the first family of constraints and the variable γ with the last constraint, is the following:

$$\begin{aligned}
\mathsf{DLP}: \quad & \min \gamma \\
& s.t. \quad \sum_{i\in N\setminus P: j\in\sigma_i} y_i x_j - \sum_{i\in N\setminus P: j\in\sigma_i^*} y_i(x_j+1) + \gamma o_j^2 \geq x_j^2, \quad \forall j \in E \\
& \qquad y_i \geq 0, \quad \forall i \in N \setminus P.
\end{aligned}$$

We get the following result.

Theorem 11.3. *For each $\lambda \in \mathbb{N} \setminus \{0\}$, $\mathsf{PoA}_{\mathcal{L}\mathcal{C}\mathcal{G}}(\lambda\text{-Cover}) \leq 1 + \frac{4\lambda+1}{4\lambda(2\lambda+1)}$.*

Proof (Sketch). For each $\lambda \in \mathbb{N} \setminus \{0\}$, set $y_i = \frac{4\lambda+1}{2\lambda+1}$ for each $i \in N \setminus P$, and $\gamma = 1 + \frac{4\lambda+1}{4\lambda(2\lambda+1)}$. With these values, the generical dual constraint becomes

$$\frac{4\lambda+1}{2\lambda+1}\left(\sum_{i \in N\setminus P: j \in \sigma_i} x_j - \sum_{i \in N\setminus P: j \in \sigma_i^*} x_j + 1\right) + \left(1 + \frac{4\lambda+1}{4\lambda(2\lambda+1)}\right) o_j^2 \geq x_j^2,$$

which, by recalling Equations (11.2) and (11.3), is non-negative whenever

$$f(x_j, o_j, s_j) := 2\lambda x_j^2 - (4\lambda+1)(x_j o_j + o_j - s_j) + \frac{8\lambda^2 + 8\lambda + 1}{4\lambda} o_j^2 \geq 0.$$

One can show that $f(x_j, o_j, s_j) \geq 0$ for any triple of non-negative integers (x_j, o_j, s_j) such that $s_j \geq \min\{\lambda, o_j\}$, thus obtaining the claim. □

We now show a matching lower bound.

Theorem 11.4. *For each* $\lambda \in \mathbb{N} \setminus \{0\}$, $\mathsf{PoA}_{\mathcal{LCG}}(\lambda\text{-Cover}) \geq 1 + \frac{4\lambda+1}{4\lambda(2\lambda+1)}$.

Proof (Sketch). For a fixed $\lambda \in \mathbb{N} \setminus \{0\}$, let n be an integer such that $n \geq 3\lambda + 1$. Consider the linear congestion game CG defined as follows.

The set of players is $N = N_1 \cup N_2$, where $|N_1| = n - \lambda$ and $|N_2| = \lambda$.

The set of resources of game CG is $E = U \cup V$, where $U = \{u_j : j \in [n-\lambda]\}$ and $V = \{v_j : j \in [n-\lambda]\}$. Each resource in U has latency function $\ell_U(x) = \frac{\lambda+1}{\lambda} x$ and each resource in V has latency function $\ell_V(x) = x$.

Each player $i \in N_1$ has two strategies: the *first strategy* $\bigcup_{j \in [\lambda]}\{u_{i+\lambda+j-1}\} \cup \bigcup_{j \in [\lambda+1]}\{v_{i+\lambda+j-1}\}$ and the *second strategy* $\bigcup_{j \in [\lambda]}\{u_{i+j-1}\} \cup \bigcup_{j \in [\lambda]}\{v_{i+j-1}\}$, where again the sums over the indexes have to be interpreted circularly. Note that, since $n - \lambda \geq 2\lambda + 1$, the first and second strategy of each player in N_1 are disjoint. Each player $i \in N_2$ has only one available strategy given by the whole set of resources E, so that N_2 is a feasible choice for λ-Cover.

One can show that the strategy profile $\boldsymbol{\sigma}$ in which each player $i \in N_1$ adopts her first strategy is a pure Nash equilibrium for $\mathsf{CG}_{\lambda\text{-Cover}}$. Now, let $\boldsymbol{\sigma}^*$ be the strategy profile in which each player in N_1 adopts her second strategy. By computing the ratio $\frac{\mathsf{SUM}(\boldsymbol{\sigma})}{\mathsf{SUM}(\boldsymbol{\sigma}^*)}$, we get the desired lower bound on $\mathsf{PoA}(\mathsf{CG}_{\lambda\text{-Cover}})$. □

In the following, we provide a lower bound holding for general affine functions.

Theorem 11.5. *For each* $\lambda \in \mathbb{N} \setminus \{0\}$ *and* $\varepsilon > 0$, *there exists an affine congestion game* CG *such that* $\mathsf{PoA}(\mathsf{CG}_{\lambda\text{-Cover}}) \geq \frac{4\lambda-1}{3\lambda-1} - \varepsilon$.

Proof. For a fixed $\lambda \in \mathbb{N} \setminus \{0\}$ and $\varepsilon > 0$, let n be an integer such that $\lambda = o(n)$. Consider the affine congestion game CG defined as follows.

The set of players is $N = N_1 \cup N_2$, where $|N_1| = n - \lambda$ and $|N_2| = \lambda$.

The set of resources is $E = U \cup V$, where $U = \{u_j : j \in [n-\lambda]\}$ and $V = \{v_j : j \in [n-\lambda]\}$. Each resource in U has latency function $\ell_U(x) = \frac{n-\lambda}{8\lambda^2-7\lambda+1}x$ and each resource in V has constant latency function $\ell_V(x) = 1$.

Each player $i \in N_1$ has two strategies: the *first strategy* $\bigcup_{j\in[3\lambda-1]}\{u_{i+\lambda+j-1}\}$ and the *second strategy* $\bigcup_{j\in[\lambda]}\{u_{i+j-1}\} \cup V$, where again the sums over the indexes have to be interpreted circularly. Note that, since $n >> \lambda$, the first and second strategy of each player in N_1 are disjoint. Each player $i \in N_2$ has only one available strategy given by the whole set of resources E, so that N_2 is a feasible choice for λ-Cover.

One can show the strategy profile $\boldsymbol{\sigma}$ in which each player $i \in N_1$ adopts her first strategy is a pure Nash equilibrium for $\mathsf{CG}_{\lambda\text{-Cover}}$. Now, let $\boldsymbol{\sigma}^*$ be the strategy profile in which each player in N_1 adopts her second strategy. By computing the ratio $\frac{\mathsf{SUM}(\boldsymbol{\sigma})}{\mathsf{SUM}(\boldsymbol{\sigma}^*)}$, one can get the desired lower bound on $\mathsf{PoA}(\mathsf{CG}_{\lambda\text{-Cover}})$. □

11.4 Price of Anarchy of Scale

For each integer $h \geq 1$, define $r_L(h) = \frac{2h^2-3}{2(h^2-1)}$ and $r_U(h) = \frac{2h^2+4h-1}{2h(h+2)}$, with $r_L(0) := 0$. Note that, since $r_L(0) = 0$, $r_U(\infty) = 1$, and $r_L(h+1) = r_U(h)$ for each $h \geq 1$, it follows that, for each $\alpha \in [0,1]$, there exists a unique integer $h := h(\alpha) \geq 1$ such that $\alpha \in [r_L(h), r_U(h)]$. We show that, for each $\alpha \in [0,1]$, $\mathsf{PoA}(\mathsf{Scale}(\alpha)) \leq 1 + \frac{(1-\alpha)(2h+1)}{(1-\alpha)h^2+\alpha h+1}$.

Also in this case we make use of the primal-dual method. To this aim, fix a linear congestion game CG and a social optimum $\boldsymbol{\sigma}^*$. For each of the $\beta := \binom{n}{\alpha n}$ possible choices of the set of coordinated players P, let $\boldsymbol{\sigma}(P)$ be a pure Nash equilibrium for the induced game. For each $j \in E$, we set $x_j(P) := n_j(\boldsymbol{\sigma}(P))$, $o_j := n_j(\boldsymbol{\sigma}^*)$ and $s_j(P) := |\{i \in P : j \in \sigma_i^*\}|$, so that $s_j(P) \leq \min\{x_j(P), o_j\}$ for each $j \in E$ and for each $P \in \mathcal{P}$, where $\mathcal{P}$ denotes the set of all the β subsets of N having cardinality αn.

We obtain the following primal linear program $\mathsf{LP} := \mathsf{LP}((\boldsymbol{\sigma}(P))_{P\in\mathcal{P}}, \boldsymbol{\sigma}^*)$:

$$\begin{aligned}
\mathsf{LP}: \quad & \max \beta^{-1} \sum_{P\in\mathcal{P}} \sum_{j\in E} a_j x_j(P)^2 \\
s.t. \quad & \sum_{j\in\sigma_i(P)} a_j x_j(P) - \sum_{j\in\sigma_i^*} a_j (x_j(P)+1) \leq 0, \quad \forall P \in \mathcal{P}, i \in N \setminus P \\
& \sum_{j\in E} a_j o_j^2 = 1 \\
& a_j \geq 0, \quad \forall j \in E.
\end{aligned}$$

The dual program $\mathsf{DLP} := \mathsf{DLP}((\boldsymbol{\sigma}(P))_{P\in\mathcal{P}}, \boldsymbol{\sigma}^*)$, obtained by associating the set of variables $(y_{i,P})_{P\in\mathcal{P}, i\in N\setminus P}$ with the first family of constraints and the variable γ with the last constraint, is the following:

$$\mathsf{DLP} \quad \min \gamma$$

$$
\begin{aligned}
s.t. \quad & \sum_{P\in\mathcal{P}}\left(\sum_{i\in N\setminus P:j\in\sigma_i(P)} y_{i,P}x_j(P) - \sum_{i\in N\setminus P:j\in\sigma_i^*} y_{i,P}(x_j(P)+1)\right) \\
& \quad + \gamma o_j^2 \geq \beta^{-1}\sum_{P\in\mathcal{P}} x_j(P)^2, \quad \forall j\in E \\
& y_{i,P}\geq 0, \quad \forall P\in\mathcal{P}, i\in N\setminus P.
\end{aligned}
$$

We get the following result.

Theorem 11.6. *For each* $\alpha\in[0,1]$, $\mathsf{PoA}(\mathsf{Scale}(\alpha))\leq 1+\frac{(1-\alpha)(2h+1)}{(1-\alpha)h^2+\alpha h+1}$.

Proof (Sketch). For any fixed value of α, set $y_{i,P}:=y\beta^{-1}$, with $y:=\frac{2h+1}{(1-\alpha)h^2+\alpha h+1}$, for each $i\in N\setminus P, P\in\mathcal{P}$, and $\gamma=1+\frac{(1-\alpha)(2h+1)}{(1-\alpha)h^2+\alpha h+1}$. With these values, the generical dual constraint becomes

$$
\begin{aligned}
& y\sum_{P\in\mathcal{P}}\beta^{-1}x_j(P)^2 - y\beta^{-1}\sum_{P\in\mathcal{P}}(x_j(P)o_j(P)+o_j(P)) + y\sum_{P\in\mathcal{P}}\beta^{-1}s_j(P)+\gamma o_j^2 \\
& \geq \sum_{P\in\mathcal{P}}\beta^{-1}x_j(P)^2,
\end{aligned}
$$

which, by using $\sum_{P\in\mathcal{P}}\beta^{-1}s_j(P)=\alpha o_j$, is equivalent to

$$
(y-1)\sum_{P\in\mathcal{P}}\beta^{-1}x_j(P)^2 - y\beta^{-1}\sum_{P\in\mathcal{P}}(x_j(P)o_j(P)+o_j(P)) + y\alpha o_j + \gamma o_j^2 \geq 0.
$$

Now note that this constraint is satisfied if

$$
f(x_j,o_j):=(y-1)x_j^2 - yo_j(x_j+1-\alpha)+\gamma o_j^2\geq 0
$$

for each pair of non-negative integers (x_j,o_j). One can show that the last inequalities is always satisfied, and this shows the claim. □

11.5 Concluding Remarks and Related Work

This chapter is based on the results obtained by Bilò and Vinci [26], who analyze the three optimal-restricted Stackelberg strategies considered in this chapter. Such strategies have been previously studied by Fotakis [88], who provides upper and lower bound on the resulting price of anarchy. By resorting to the primal-dual method, Bilò and Vinci [26] either tighten or improve most of the existing bounds on the price of anarchy, which are described in the following.

- Regarding Largest Latency First, Fotakis [88] shows that $\mathsf{PoA}(\mathsf{LLF}(\alpha))\leq \min\left\{\frac{20-11\alpha}{8},\frac{3-2\alpha+\sqrt{5-4\alpha}}{2}\right\}$ and $\mathsf{PoA}(\mathsf{LLF}(\alpha))\geq\frac{5(2-\alpha)}{4+\alpha}$, where the lower bound holds even for the restricted case of symmetric players. The results presented

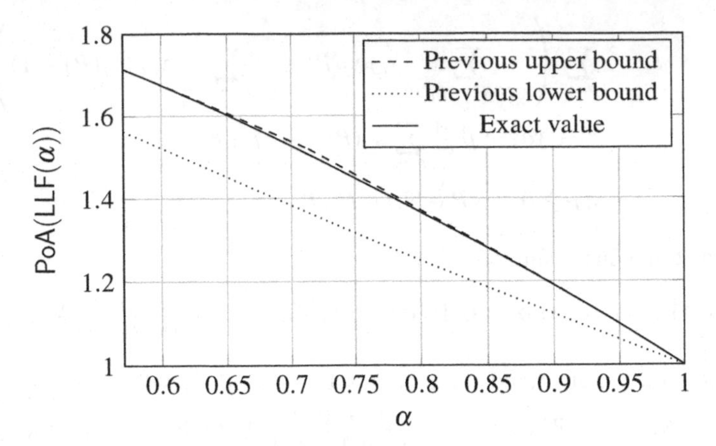

Fig. 11.1 Comparison between the exact value of $\mathsf{PoA}(\mathsf{LLF}(\alpha))$ and the upper and lower bounds previously known in the literature for the case of $\alpha \in [4/7, 1]$.

in Section 11.2 show how to close this gap for non-symmetric players by proving that $\mathsf{PoA}(\mathsf{LLF}(\alpha)) = \frac{20-11\alpha}{8}$ for $\alpha \in [0, 4/7]$[2] and $\mathsf{PoA}(\mathsf{LLF}(\alpha)) = \frac{4-3\alpha+\sqrt{4\alpha-3\alpha^2}}{2}$ for $\alpha \in [4/7, 1]$.
For a quantitative comparison of the values in this case, see Figure 11.1.

- For λ-Cover, Fotakis [88] shows that $\mathsf{PoA}_{\mathcal{AEG}}(\lambda\text{-Cover}) \leq \frac{4\lambda-1}{3\lambda-1}$ and that $\mathsf{PoA}_{\mathcal{LEG}}(\lambda\text{-Cover}) \leq 1 + \frac{1}{2\lambda}$, while no lower bounds are provided. In this case, the results of Section 11.3 provide the exact value of $\mathsf{PoA}_{\mathcal{LEG}}(\lambda\text{-Cover})$ by improving the upper bound previously given by Fotakis [88] and providing matching lower bounding instances whereas no trivial lower bounds were previously known. Furthermore, the results on affine latency functions show that the upper bound given by Fotakis [88] is asymptotically tight.
For a quantitative comparison of the values, see Figure 11.2.
- Relatively to Scale, Fotakis [88] shows that $\mathsf{PoA}(\mathsf{Scale}(\alpha)) \leq \max\left\{\frac{5-3\alpha}{2}, \frac{5-4\alpha}{3-2\alpha}\right\}$ and $\mathsf{PoA}(\mathsf{Scale}(\alpha)) \geq \frac{2}{1+\alpha}$, where the lower bound holds for any randomized optimal-restricted Stackelberg strategy even when applied to symmetric players. In Section 11.4, we provide better upper bounds with respect to those provided by Fotakis [88] for the case of $\alpha \in (5/6, 1)$[3]; for a quantitative comparison of the values, see Figure 11.3.

The work of Fotakis [88] is the first work on Stackelberg strategies that considers atomic congestion games. Efficient Stackelberg strategies have been mainly studied within the context of *non-atomic* congestion games, that is, the case in which there are infinitely many players so that each player contributes for a negligible

[2] We observe that the upper bound of $\frac{20-11\alpha}{8}$ was already established by Fotakis [88].

[3] For $h = 1$, which covers the case $\alpha \in [0, 5/6]$, and for $h = \infty$, which covers the case $\alpha = 1$, the upper bounds of Section 11.4 coincide with the ones given by Fotakis [88], while, in all the other cases, they are better.

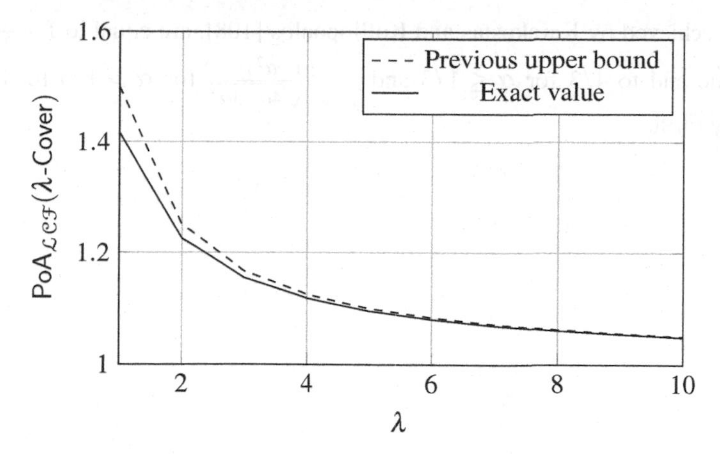

Fig. 11.2 Comparison between the exact value of $\mathsf{PoA}_{\mathcal{LCG}}(\lambda\text{-Cover})$ and the upper bound previously known in the literature.

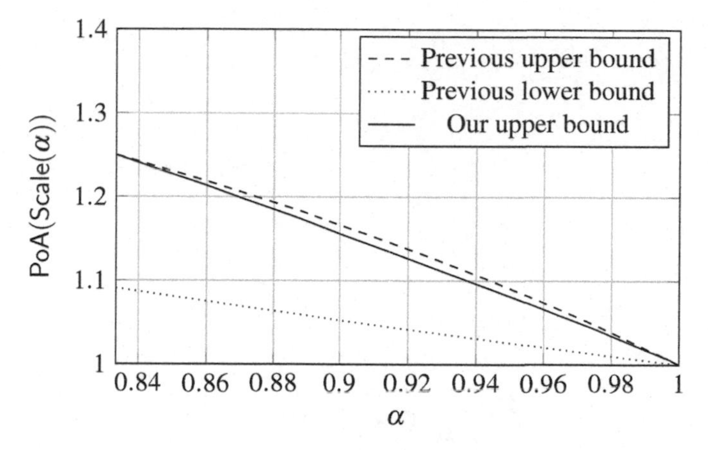

Fig. 11.3 Comparison between our upper bounds on $\mathsf{PoA}(\mathsf{Scale}(\alpha))$ and the upper and lower bounds previously known in the literature for the case of $\alpha \in [5/6, 1]$.

fraction to the resources congestion. Roughgarden [156] shows that it is NP-hard to compute an optimal Stackelberg strategy even in games played on parallel link networks. Under this restriction, an approximation scheme has been given by Kumar and Marathe [115] for the case of polynomial latency functions. The works of Roughgarden [156], [171], Correa and Stier Moses [78], and Karakostas and Kolliopoulos [108] investigate the price of anarchy achieved by two natural Stackelberg strategies, namely Largest Latency First and Scale, as a function of the fraction of the coordinated players, denoted as α. These two strategies, introduced by Roughgarden [156], are defined similarly as in atomic games: Largest Latency First assigns the coordinated players to the largest cost strategies in the social optimum, while Scale simply employs the social optimum scaled by α. The best known upper bounds, achieved by Karakostas and Kolliopoulos [108], are equal to $1 + \frac{1-\alpha}{2\sqrt{1-\alpha}+1}$ for Scale and to $4/3$ for $\alpha \leq 1/3$ and $\frac{2(1-\alpha^2)}{2-\alpha-\sqrt{4\alpha-3\alpha^2}}$ for $\alpha > 1/3$ for Largest Latency First.

Appendix

Appendix A
Other Strategies to Improve the Efficiency of Selfish Outcomes

In Chapters 8–11, we considered the problem of improving the efficiency of selfish outcomes by means of either taxation mechanisms or Stackelberg strategies. In this chapter, we briefly describe two other mechanisms adopted to reduce the price of anarchy in resource selection games: *coordination mechanisms* and *cost-sharing protocols*. We provide the main definitions and the related work, so that the interested reader can deepen the state-of-the-art on these mechanisms and complement the results provided for taxation mechanisms and Stackelberg strategies.

A.1 Coordination Mechanisms

A well-known class of optimization problems that is strictly connected with weighted congestion games and, in particular, with linear load balancing games with unrelated machines (Section 2.6), is that of *machine scheduling problems*. Similarly as in load balancing problems, in more general machine scheduling problems we have a set of machines and a set of jobs that can be assigned to the machines, but each machine is equipped with a *local scheduling policy* that runs in a decentralized fashion and decides how to schedule its jobs with the aim of minimizing a certain objective function (e.g., the makespan, the weighted total completion time, etc.). For their wide applicability to several practical problems (e.g., CPU scheduling, queuing optimization in routing problems, etc.), machine scheduling models have been constituting a central topic in computer science and operation research for several decades.

With the advent of algorithm game theory, machine scheduling models have been also studied from a game-theoretic perspective: each job is a selfish player and chooses a machine with the aim of minimizing her individual processing time, regardless of the other players' processing times. To reduce as much as possible the impact of selfish behavior (as usual, measured via the price of anarchy), we can decide which kind of local scheduling policy must be adopted by the machines. A set of such local scheduling policies, one for each machine, is called *coordination mechanism*.

V. Bilò, C. Vinci, *Coping with Selfishness in Congestion Games*, Monographs in Theoretical Computer Science. An EATCS Series, https://doi.org/10.1007/978-3-031-30261-9

Since the definition of coordination mechanisms [68], a long list of papers has mostly focused on the design of policies in machine scheduling settings, aiming to minimize the makespan (e.g., [1, 14, 19, 49, 50, 106, 111, 153]) or the total (weighted) completion time (e.g., [2, 72, 77, 105, 153]). Apart from machine scheduling settings, coordination mechanisms have also been proposed for congestion games [62]. A class of strategies that exhibits several similarities with coordination mechanisms is that of taxation mechanisms (see Chapter 8), where, instead of a local scheduling policy, we have a local tax that is charged to the players selecting a resource.

A.2 Cost-sharing Protocols

A class of strategies that is strictly connected with coordination mechanisms and has not been considered in the previous chapters is that of *cost-sharing games* [8, 57, 64, 96, 97, 112, 123, 129, 130, 176]. In cost-sharing games we have a set of resources, and the cost of each resource depends on the set of players selecting them. A *cost-sharing protocol*, similarly to a local scheduling policy, decides how to share the cost of each resource among the players selecting it, again, with the aim of guiding players toward more efficient equilibria. Two of the most well-known cost-sharing protocols are the *marginal contribution* and the *Shapley value* [166].

Caragiannis et al. [56] studied the interplay between coordination mechanisms and scheduling algorithms in more depth, and outlined interesting connections between coordination mechanisms and cost-sharing policies. Such connections are exploited to improve the analysis of existing coordination mechanisms for the weighted total completion time. Furthermore, they designed a coordination mechanisms that can be used to obtain a 1.81-approximation algorithm for the underlying optimization problem of minimizing the total weighted completion time (in a centralized fashion); to the best of our knowledge, this is the best approximation guarantee among combinatorial polynomial-time algorithms for the considered optimization problem. Anyway, when considering more general non-combinatorial approaches (e.g., LP or convex relaxations), the best-approximation factor for the problem of minimizing the the total weighted completion time was obtained by Bansal et al. [15] and Li [119], who provided a $(1.5-c)$-approximation algorithm (for a small positive constant c) and improved the 1.5-approximation previously achieved by Sethuraman and Squillante [165], Skutella [169][1].

[1] The problem of scheduling jobs on unrelated machines aiming to minimize the total weighted completion time has been studied extensively in the machine scheduling literature (for a detailed list of some of the classic results see [117, Chapter 11]).

Appendix B
Miscellaneous

In this chapter of the appendix, we provide further notions/results that are strictly related with the content of the book.

B.1 Duality Theory in Linear Programming

In this section, we give a brief overview of the duality theory in linear programming, and we provide the main theorems which the primal-dual method (Chapter 3.1) is based on. For further details, see Matouek and Gärtner [124].

Primal-Dual Pair

Let us consider the following linear program in n variables $x_1, \ldots, x_n$ and m constraints (excluding the constraints imposing the non-negativity of the variables):

$$\begin{aligned}
\mathsf{LP}: \quad \max \quad & c_1x_1 + c_2x_2 + \ldots + c_nx_n \\
s.t. \quad & a_{1,1}x_1 + a_{1,2}x_2 + \ldots + a_{1,n}x_n \leq b_1 \\
& a_{2,1}x_1 + a_{2,2}x_2 + \ldots + a_{2,n}x_n \leq b_2 \\
& \vdots \\
& a_{m,1}x_1 + a_{m,2}x_2 + \ldots + a_{m,n}x_n \leq b_m \\
& x_i \geq 0, \quad \forall i \in [n],
\end{aligned}$$

where $(a_{i,j})_{i,j} \in \mathbb{R}^{nm}$ and $(c_i)_i \in R^n$ are some fixed real coefficients. The *dual program* of LP is a linear program in m variables $y_1, \ldots, y_m$ and n constraints:

$$\mathsf{DLP}: \quad \min \quad b_1y_1 + b_2y_2 + \ldots + b_my_m$$

V. Bilò, C. Vinci, *Coping with Selfishness in Congestion Games*, Monographs in Theoretical Computer Science. An EATCS Series, https://doi.org/10.1007/978-3-031-30261-9

$$
\begin{aligned}
s.t.\quad & a_{1,1}y_1 + a_{2,1}y_2 + \ldots + a_{m,1}y_n \geq c_1 \\
& a_{1,2}y_1 + a_{2,2}y_2 + \ldots + a_{m,2}y_n \geq c_2 \\
& \vdots \\
& a_{1,n}y_1 + a_{2,n}y_2 + \ldots + a_{m,n}y_n \geq c_n \\
& y_j \geq 0, \quad \forall j \in [m].
\end{aligned}
$$

Each variable y_j of the dual is associated with the jth constraint of the primal, and each variable x_i of the primal is associated with the ith constraint of the dual. The dual program DLP can be transformed into an equivalent optimization problem by multiplying all the coefficients by -1, and by replacing "$\geq$" and "min" with "$\leq$" and "max", respectively. If we consider such equivalent representation of DLP, and we compute the dual of DLP, we obtain a linear program that is equivalent to the initial linear program LP. The initial linear program LP is called *primal program*, and the pair (LP, DLP) is called *primal-dual* pair.

Main Theorems

In the following, we provide the main theorems that relate optimal and feasible solutions of the primal program to those of the corresponding dual. We say that a linear program is *feasible* if it admits a solution and it is *bounded* if its optimal value is finite.

Theorem B.1 (Weak Duality Theorem). *Let* $\mathbf{x} = (x_1, \ldots, x_n)$ *and* $\mathbf{y} = (y_1, \ldots, y_m)$ *be two feasible solutions of* LP *and* DLP*, respectively. Then* $\sum_{i=1}^{n} c_i x_i \leq \sum_{j=1}^{m} b_j y_j$*, i.e., the value of any feasible solution of* LP *is a lower bound on the value of any feasible solution of* DLP*. Furthermore, if the optimal value of* LP *is unbounded then* DLP *is unfeasible.*

Theorem B.2 (Strong Duality Theorem). *Assume that* LP *is feasible and bounded, and let* $\mathbf{x}^* = (x_1^*, \ldots, x_n^*)$ *be an optimal solution. Then* LP *is feasible and bounded too, and* $\sum_{i=1}^{n} c_i x_i^* = \sum_{j=1}^{m} b_j y_j^*$*, where* $\mathbf{y}^* = (y_1^*, \ldots, y_m^*)$ *is an optimal solution of* DLP.

Theorem B.3 (Complementary Slackness Conditions). *Let* $\mathbf{x}^* = (x_1^*, \ldots, x_n^*)$ *and* $\mathbf{y}^* = (y_1^*, \ldots, n_n^*)$ *be two feasible solutions of the primal and the dual program, respectively. Then* $\mathbf{x}^*$ *and* $\mathbf{y}^*$ *are respectively optimal solutions for the primal and the dual if and only if the following complementary slackness conditions hold:*

1. *for any* $i \in [n]$*, either* $x_i^* = 0$*, or* $\sum_{j=1}^{m} a_{j,i} y_j = c_i$ *(i.e., the dual constraint associated with the ith primal variable is tight), or both equalities hold;*
2. *for any* $j \in [m]$*, if* $y_j^* = 0$*, or* $\sum_{i=1}^{n} a_{i,j} x_i = b_j$ *(i.e., the primal constraint associated with the jth dual variable is tight), or both equalities hold.*

B.2 Smoothness Framework

The smoothness analysis (Roughgarden [158]) has been introduced to give almost tight upper bounds on the price of anarchy of several games, and applies successfully to the case of congestion games. In particular, an atomic congestion game CG is (λ,μ)-*smooth* if, for any pair of strategy profiles $\boldsymbol{\sigma},\boldsymbol{\sigma}^*$ of CG, the following condition holds:

$$\sum_{i\in N} cost_i(\boldsymbol{\sigma}_{-i},\boldsymbol{\sigma}_i^*) \leq \lambda\cdot\mathsf{SUM}(\boldsymbol{\sigma}^*)+\mu\cdot\mathsf{SUM}(\boldsymbol{\sigma}). \tag{B.1}$$

A class $\mathcal{G}$ of congestions games is (λ,μ)-smooth if any game $\mathsf{CG}\in\mathcal{G}$ is (λ,μ)-smooth. The following results give a characterization of the price of anarchy of atomic congestion games with respect to the social function SUM.

Theorem B.4 (Roughgarden [158]). *(i) Given a (λ,μ)-smooth class $\mathcal{G}$ of (weighted or unweighted) congestion games with $\mu\in[0,1]$ and $\lambda\geq 0$, then $\mathsf{PoA}(\mathcal{G})\leq\frac{\lambda}{1-\mu}$. (ii) If $\mathcal{G}=\mathsf{U}(\mathcal{C})$ for some class $\mathcal{C}$ of latency functions, we have the following characterization of the price of anarchy:*

$$\begin{aligned}
&\mathsf{PoA}(\mathcal{G})\\
&=\inf\left\{\frac{\lambda}{1-\mu}:\mathcal{G}\textit{ is }(\lambda,\mu)\textit{-smooth}\right\}\\
&=\inf_{\mu\in[0,1),\lambda\geq 0}\left\{\frac{\lambda}{1-\mu}:of(k+1)\leq\lambda of(o)+\mu kf(k),\forall f\in\mathcal{C},k\in\mathbb{Z}_{\geq 0},o\in\mathbb{Z}_{\geq 0}\right\}.
\end{aligned} \tag{B.2}$$

One can easily see that the PoA bound for (exact) Nash equilibria provided in Theorem 4.6 (Chapter 4) for unweighted games and that of Theorem B.4.ii are equivalent. In particular, the infimum of $\frac{\lambda}{1-\mu}$ considered in (B.2) is equal to the value $\gamma_{\mathsf{U}}(\mathsf{PoA}_0,\mathcal{G})$ used in Theorem 4.6. We also stress that the approach considered in [158] to show the tight bound provided in (B.2) resorts to a technical lemma similar to Lemma 4.4 (Chapter 4).

A tight characterization of the price of anarchy for weighted congestion games holds under mild assumptions on the considered latency functions.

Theorem B.5 (Bhawalkar et al. [20]). *Given a class of weighted congestion games $\mathcal{G}=\mathsf{W}(\mathcal{C})$, where $\mathcal{C}$ is closed under abscissa scaling, the following characterization of the price of anarchy holds:*

$$\begin{aligned}
&\mathsf{PoA}(\mathcal{G})\\
&=\inf\left\{\frac{\lambda}{1-\mu}:\mathcal{G}\textit{ is }(\lambda,\mu)\textit{-smooth}\right\}\\
&=\inf_{\mu\in[0,1),\lambda\geq 0}\left\{\frac{\lambda}{1-\mu}:of(k+o)\leq\lambda of(o)+\mu kf(k),\forall f\in\mathcal{C},k\geq 0,o\geq 0\right\}
\end{aligned} \tag{B.3}$$

Similarly to the unweighted case, one can easily see that the PoA bound for (exact) Nash equilibria provided in Theorem 4.1 (Chapter 4) for weighted games and that of Theorem B.5 are equivalent. We also stress that the approach considered in [20] to show the tight bound provided in (B.3) resorts to a technical lemma similar to Lemma 4.1.

References

[1] F. Abed and C.-C. Huang. Preemptive coordination mechanisms for unrelated machines. In *Proceedings of the 20th Annual European Sympoisum on Algorithms (ESA)*, pages 12–23, 2012.

[2] F. Abed, J. R. Correa, and C.-C. Huang. Optimal coordination mechanisms for multi-job scheduling games. In *Proceedings of the 22nd Annual European Sympoisum on Algorithms (ESA)*, pages 13–24, 2014.

[3] H. Ackermann, H. Röglin, and B. Vöcking. On the impact of combinatorial structure on congestion games. *Journal of ACM*, 55(6), 2008.

[4] H. Ackermann, H. Röglin, and B. Vöcking. Pure Nash equilibria in player-specific and weighted congestion games. *Theoretical Computer Science*, 410 (17):1552–1563, 2009.

[5] M. Aghassi and D. Bertsimas. Robust game theory. *Mathematical Programming*, 107(1-2):231–273, 2006.

[6] R. K. Ahuja, T. L. Magnanti, and J. B. Orlin. *Network Flows: Theory, Algorithms, and Applications*. Pearson, 1993.

[7] S. Aland, D. Dumrauf, M. Gairing, B. Monien, and F. Schoppmann. Exact price of anarchy for polynomial congestion games. *SIAM Journal on Computing*, 40(5):1211–1233, 2011.

[8] E. Anshelevich, A. Dasgupta, J. M. Kleinberg, É. Tardos, T. Wexler, and T. Roughgarden. The price of stability for network design with fair cost allocation. *SIAM Journal on Computing*, 38(4):1602–1623, 2008.

[9] I. Ashlagi, D. Monderer, and M. Tennenholtz. Two-terminal routing games with unknown active players. *Artificial Intelligence*, 173(15):1441–1455, 2009.

[10] R. Aumann. Subjectivity and correlation in randomized strategies. *Journal of Mathematical Economics*, 1(1):67–96, 1974.

[11] B. Awerbuch, Y. Azar, E. F. Grove, M.-Y. Kao, P. Krishnan, and Vitter J. S. Load balancing in the l_p norm. In *Proceedings of the 36th Annual Symposium on Foundations of Computer Science (FOCS)*, pages 383–391, 1995.

V. Bilò, C. Vinci, *Coping with Selfishness in Congestion Games*, Monographs in Theoretical Computer Science. An EATCS Series, https://doi.org/10.1007/978-3-031-30261-9

[12] B. Awerbuch, Y. Azar, and A. Epstein. The price of routing unsplittable flow. In *Proceedings of the 37th Annual ACM Symposium on Theory of Computing (STOC)*, pages 57–66, 2005.

[13] B. Awerbuch, Y. Azar, A. Epstein, V. S. Mirrokni, and A. Skopalik. Fast convergence to nearly optimal solutions in potential games. In *Proceedings of the 9th ACM Conference on Electronic Commerce (EC)*, pages 264–273, 2008.

[14] Y. Azar, L. Fleischer, K. Jain, V. S. Mirrokni, and Z. Svitkina. Optimal coordination mechanisms for unrelated machine scheduling. *Operations Research*, 63(3):489–500, 2015.

[15] N. Bansal, A. Srinivasan, and O. Svensson. Lift-and-round to improve weighted completion time on unrelated machines. In *Proceedings of the 48th Annual ACM Symposium on Theory of Computing (STOC)*, pages 156–167, 2016.

[16] Y. Bartal, A. Fiat, H. Karloff, and R. Vohra. New algorithms for an ancient scheduling problem. *Journal of Computer and System Sciences*, 51(3):359 – 366, 1995.

[17] M. J. Beckmann, C. B. McGuire, and C. B. Winsten. *Studies in the Economics of Transportation*. Yale University Press, 1956.

[18] F. Benita, V. Bilò, B. Monnot, G. Piliouras, and C. Vinci. Data-driven models of selfish routing: Why price of anarchy does depend on network topology. In *Proceedings of the 16th International Conference on Web and Internet Economics (WINE)*, pages 252–265, 2020.

[19] S. Bhattacharya, S. Im, J. Kulkarni, and K. Munagala. Coordination mechanisms from (almost) all scheduling policies. In *Proceedings of the 5th Conference on Innovations in Theoretical Computer Science (ITCS)*, pages 121–134, 2014.

[20] K. Bhawalkar, M. Gairing, and T. Roughgarden. Weighted congestion games: price of anarchy, universal worst-case examples, and tightness. *ACM Transactions on Economics and Computation*, 2(4):1–23, 2014.

[21] V. Bilò. On linear congestion games with altruistic social context. In *Proceedings of the 20th International Conference on Computing and Combinatorics (COCOON)*, pages 547–558, 2014.

[22] V. Bilò. A unifying tool for bounding the quality of non-cooperative solutions in weighted congestion games. *Theory of Computing Systems*, 62(5):1288–1317, 2018.

[23] V. Bilò and M. Paladini. On the performance of mildly greedy players in cut games. *Journal of Combinatorial Optimization*, 32(4):1036–1051, 2016.

[24] V. Bilò and C. Vinci. On the impact of singleton strategies in congestion games. In *Proceedings of the 25th Annual European Symposium on Algorithms (ESA)*, pages 17:1–17:14, 2017.

[25] V. Bilò and C. Vinci. On the impact of singleton strategies in congestion games (extended abstract). In *CEUR Workshop Proceedings*, volume 1949, pages 223–227, 2017.

[26] V. Bilò and C. Vinci. On stackelberg strategies in affine congestion games. *Theory of Computing Systems*, 63(6):1228–1249, 2019.
[27] V. Bilò and C. Vinci. Dynamic taxes for polynomial congestion games. *ACM Transantions on Economics and Computation*, 7(3):15:1–15:36, 2019.
[28] V. Bilò and C. Vinci. Congestion games with priority-based scheduling. In *Proceedings of the 13th International Symposium on Algorithmic Game Theory (SAGT)*, pages 67–82, 2020.
[29] V. Bilò and C. Vinci. The price of anarchy of affine congestion games with similar strategies. *Theoretical Computer Science*, 806:641–654, 2020.
[30] V. Bilò and C. Vinci. The impact of selfish behavior in load balancing games. *CoRR abs/2202.12173*, 2022.
[31] V. Bilò, A. Fanelli, M. Flammini, and L. Moscardelli. Performances of one-round walks in linear congestion games. *Theory of Computing Systems*, 49 (1):24–45, 2011.
[32] V. Bilò, I. Caragiannis, A. Fanelli, and G. Monaco. Improved lower bounds on the price of stability of undirected network design games. *Theory of Computing Systems*, 52(4):668–686, 2013.
[33] V. Bilò, M. Flammini, G. Monaco, and L. Moscardelli. Some anomalies of farsighted strategic behavior. *Theory of Computing Systems*, 56(1):156–180, 2015.
[34] V. Bilò, A. Fanelli, and L. Moscardelli. On lookahead equilibria in congestion games. *Mathematical Structures in Computer Science*, 27(2):197–214, 2017.
[35] V. Bilò, L. Moscardelli, and C. Vinci. Uniform mixed equilibria in network congestion games with link failures. In *Proceedings of the 45th International Colloquium on Automata, Languages and Programming (ICALP)*, pages 146:1–146:14, 2018.
[36] V. Bilò, L. Gourvès, and J. Monnot. Project games. In *Proceedings of the 11th International Conference on Algorithms and Complexity (CIAC)*, pages 75–86, 2019.
[37] V. Bilò, M. Flammini, V. Gallotti, and C. Vinci. On multidimensional congestion games. *Algorithms*, 13(10):261, 2020.
[38] V. Bilò, M. Flammini, and L. Moscardelli. The price of stability for undirected broadcast network design with fair cost allocation is constant. *Games and Economic Behavior*, 123:359–376, 2020.
[39] V. Bilò, G. Monaco, L. Moscardelli, and C. Vinci. Nash social welfare in selfish and online load balancing. In *Proceedings of the 16th International Conference on Web and Internet Economics (WINE)*, pages 323–337, 2020.
[40] V. Bilò, D. Ferraioli, and C. Vinci. General opinion formation games with social group membership. In *Proceedings of the 31st International Joint Conference on Artificial Intelligence (IJCAI)*, pages 88–94, 2022.
[41] V. Bilò, G. Monaco, L. Moscardelli, and C. Vinci. Nash social welfare in selfish and online load balancing. *ACM Trans. Econ. Comput.*, 10(2), 2022.
[42] Vittorio Bilò. On the robustness of the approximate price of anarchy in generalized congestion games. *Theoretical Computer Science*, 906:94–113, 2022.

[43] S. Boyd and L. Vandenberghe. *Convex optimization.* Cambridge University Press, 2004.

[44] P. N. Brown and J. R. Marden. Optimal mechanisms for robust coordination in congestion games. In *Proceedings of the 54th IEEE Conference on Decision and Control (CDC)*, pages 2283–2288, 2015.

[45] P. N. Brown and J. R. Marden. The robustness of marginal-cost taxes in affine congestion games. *IEEE Transactions on Automatic Control*, 62(8): 3999–4004, 2017.

[46] P. N. Brown and J. R. Marden. The benefit of perversity in taxation mechanisms for distributed routing. In *Proceedings of the 56th IEEE Annual Conference on Decision and Control (CDC)*, pages 6229–6234, 2017.

[47] N. Buckbinder and J. Naor. The design of competitive online algorithms via a primal–dual approach. *Foundations and Trends in Theoretical Computer Science*, 3(2–3):93–263, 2009.

[48] I. Caragiannis. Better bounds for online load balancing on unrelated machines. In *Proceedings of the ACM-SIAM Symposium on Discrete Algorithms (SODA)*, pages 972–981, 2008.

[49] I. Caragiannis. Efficient coordination mechanisms for unrelated machine scheduling. *Algorithmica*, 66(3):512–540, 2013.

[50] I. Caragiannis and A. Fanelli. An almost ideal coordination mechanism for unrelated machine scheduling. *Theory of Computing Systems*, 63(1):114–127, 2019.

[51] I. Caragiannis and A. Fanelli. On approximate pure nash equilibria in weighted congestion games with polynomial latencies. *Journal of Computer and System Sciences*, 117:40–48, 2021.

[52] I. Caragiannis, C. Kaklamanis, and P. Kanellopoulos. Taxes for linear atomic congestion games. *ACM Transactions on Algorithms*, 7(1):13:1–13:31, 2010.

[53] I. Caragiannis, A. Fanelli, N. Gravin, and A. Skopalik. Efficient computation of approximate pure nash equilibria in congestion games. In *Proceedings of the 52nd Annual Symposium on Foundations of Computer Science (FOCS)*, pages 532–541, 2011.

[54] I. Caragiannis, M. Flammini, C. Kaklamanis, P. Kanellopoulos, and L. Moscardelli. Tight bounds for selfish and greedy load balancing. *Algorithmica*, 61(3):606–637, 2011.

[55] I. Caragiannis, A. Fanelli, N. Gravin, and A. Skopalik. Approximate pure Nash equilibria in weighted congestion games: existence, efficient computation and structure. *ACM Transactions on Economics and Computation*, 3(1), 2015.

[56] I. Caragiannis, V. Gkatzelis, and C. Vinci. Coordination mechanisms, cost-sharing, and approximation algorithms for scheduling. In *Proceedings of the 13th International Conference on Web and Internet Economics (WINE)*, pages 74–87, 2017.

[57] H.-L. Chen, T. Roughgarden, and G. Valiant. Designing network protocols for good equilibria. *SIAM Journal on Computing*, 39(5):1799–1832, 2010.

[58] P.-A. Chen, B. de Keijzer, D. Kempe, and G. Schäfer. The robust price of anarchy of altruistic games. In *Proceedings of the 7th International Conference on Internet and Network Economics (WINE)*, pages 383–390, 2011.
[59] G. Christodoulou and E. Koutsoupias. The price of anarchy of finite congestion games. In *Proceedings of the 37th Annual ACM Symposium on Theory of Computing (STOC)*, pages 67–73, 2005.
[60] G. Christodoulou and E. Koutsoupias. On the price of anarchy and stability of correlated equilibria of linear congestion games. In *Proceedings of the 13th Annual European Symposium on Algorithms (ESA)*, pages 59–70, 2005.
[61] G. Christodoulou, E. Koutsoupias, and P. G. Spirakis. On the performance of approximate equilibria in congestion games. *Algorithmica*, 61(1):116–140, 2011.
[62] G. Christodoulou, K. Mehlhorn, and E. Pyrga. Improving the price of anarchy for selfish routing via coordination mechanisms. In *Proceedings of the 19th European Conference on Algorithms (ESA)*, pages 119–130, 2011.
[63] G. Christodoulou, V. S. Mirrokni, and A. Sidiropoulos. Convergence and approximation in potential games. *Theoretical Computer Science*, 438:13–27, 2012.
[64] G. Christodoulou, V. Gkatzelis, and A. Sgouritsa. Cost-sharing methods for scheduling games under uncertainty. In *Proceedings of the 18th ACM Conference on Economics and Computation (EC)*, pages 441–458, 2017.
[65] G. Christodoulou, M. Gairing, Y. Giannakopoulos, and P. G. Spirakis. The price of stability of weighted congestion games. *SIAM Journal on Computing*, 48(5):1544–1582, 2019.
[66] G. Christodoulou, M. Gairing, Y. Giannakopoulos, D. Poças, and C. Waldmann. Existence and complexity of approximate equilibria in weighted congestion games. In *Proceedings of the 47th International Colloquium on Automata, Languages, and Programming (ICALP)*, pages 32:1–32:18, 2020.
[67] George Christodoulou and Martin Gairing. Price of stability in polynomial congestion games. *ACM Transactions on Economics and Computation*, 4(2): 10:1–10:17, 2016.
[68] George Christodoulou, Elias Koutsoupias, and Akash Nanavati. Coordination mechanisms. *Theoretical Computer Science*, 410(36):3327–3336, 2009.
[69] S. J. Chung and K. G. Murty. Polynomially bounded ellipsoid algorithms for convex quadratic programming. In *Nonlinear Programming 4*, pages 439 – 485. 1981.
[70] R. Cole, Y. Dodis, and T. Roughgarden. Pricing network edges for heterogeneous selfish users. In *Proceedings of the 35th Annual ACM Symposium on Theory of Computing (STOC)*, pages 521–530, 2003.
[71] R. Cole, Y. Dodis, and T. Roughgarden. How much can taxes help selfish routing? *Journal of Computer and System Sciences*, 72(3):444–467, 2006.
[72] R. Cole, J.R. Correa, V. Gkatzelis, V. Mirrokni, and N. Olver. Decentralized utilitarian mechanisms for scheduling games. *Games and Economic Behavior*, 92:306–326, 2014.

[73] R. Colini-Baldeschi, M. Klimm, and M. Scarsini. Demand-independent optimal tolls. In *Proceedings of the 45th International Colloquium on Automata, Languages, and Programming (ICALP)*, pages 151:1–151:14, 2018.

[74] R. Cominetti, J. R. Correa, and N. E. Stier Moses. The impact of oligopolistic competition in networks. *Operations Research*, 57(6):1421–1437, 2009.

[75] R. Cominetti, M. Scarsini, M. Schröder, and N. E. Stier-Moses. Price of anarchy in stochastic atomic congestion games with affine costs. In *Proceedings of the 20th ACM Conference on Economics and Computation (EC)*, pages 579–580, 2019.

[76] J. Correa, J. de Jong, B. de Keijzer, and M. Uetz. The curse of sequentiality in routing games. In *Proceedings of the 11th International Conference on Web and Internet Economics (WINE)*, pages 258–271, 2015.

[77] J. R. Correa and M. Queyranne. Efficiency of equilibria in restricted uniform machine scheduling with total weighted completion time as social cost. *Naval Research Logistics (NRL)*, 59(5):384–395, 2012.

[78] J. R. Correa and N. E. Stier Moses. Stackelberg routing in atomic network games. 2007. Technical Report.

[79] J. R. Correa, A. S. Schulz, and N. E Stier Moses. Selfish routing in capacitated networks. *Mathematics of Operations Research*, 29(4):961–976, 2004.

[80] J. de Jong and M. Uetz. The sequential price of anarchy for affine congestion games with few players. *Operations Research Letters*, 47(2):133–139, 2019.

[81] J. de Jong, M. Klimm, and M. Uetz. Efficiency of equilibria in uniform matroid congestion games. In *Proceedings of the 9th International Symposium on Algorithmic Game Theory (SAGT)*, pages 105–116, 2016.

[82] C. Deeparnab, M. Aranyak, and N. Viswanath. Fairness and optimality in congestion games. In *Proceedings of the 6th ACM Conference on Electronic Commerce (EC)*, pages 52–57, 2005.

[83] A. Fabrikant, C. H. Papadimitriou, and K. Talwar. The complexity of pure Nash equilibria. In *Proceedings of the 36th Annual ACM Symposium on Theory of Computing (STOC)*, pages 604–612, 2004.

[84] B. L. Ferguson, P. N. Brown, and J. R. Marden. Carrots or sticks? the effectiveness of subsidies and tolls in congestion games. In *Proceedings of the 2020 American Control Conference (ACC)*, pages 1853–1858, 2020.

[85] B. L. Ferguson, P. N. Brown, and J. R. Marden. The effectiveness of subsidies and taxes in atomic congestion games. *IEEE Control Systems Letters*, 6:614–619, 2022.

[86] A. Fiat, K. Kaplan, M. Levy, S. Olonetsky, and R. Shabo. On the price of stability for designing undirected networks with fair cost allocations. In *Proceedings of the 33rd International Colloquium on Automata, Languages and Programming (ICALP)*, pages 608–618, 2006.

[87] L. Fleischer, K. Jain, and M. Mahdian. Tolls for heterogeneous selfish users in multicommodity networks and generalized congestion games. In *Proceedings of the 45th Annual IEEE Symposium on Foundations of Computer Science (FOCS)*, pages 277–285, 2004.

[88] D. Fotakis. Stackelberg strategies for atomic congestion games. *Theory of Computing Systems*, 47(1):218–249, 2010.

[89] D. Fotakis and P. G. Spirakis. Cost-balancing tolls for atomic network congestion games. In *Proceedings of the 3rd International Conference on Internet and Network Economics (WINE)*, pages 179–190, 2007.

[90] D. Fotakis, S. Kontogiannis, and P. Spirakis. Selfish unsplittable flows. *Theoretical Computer Science*, 348:226–239, 2005.

[91] D. Fotakis, S. C. Kontogiannis, and P. G. Spirakis. Symmetry in network congestion games: pure equilibria and anarchy cost. In *Proceedings of the 3rd Workshop on Approximation and Online Algorithms (WAOA)*, pages 161–175, 2005.

[92] M. Gairing and F. Schoppmann. Total latency in singleton congestion games. In *Proceedings of the Third International Workshop on Internet and Network Economics (WINE)*, pages 381–387, 2007.

[93] M. Gairing, T. Lücking, M. Mavronicolas, and B. Monien. The price of anarchy for polynomial social cost. *Theoretical Computer Science*, 369(1–3):116–135, 2006.

[94] M Gairing, T. Lücking, M. Mavronicolas, B. Monien, and M. Rode. Nash equilibria in discrete routing games with convex latency functions. *Journal of Computer and System Sciences*, 74(7):1199–1225, 2008.

[95] Y. Giannakopoulos, G. Noarov, and A. S. Schulz. Computing approximate equilibria in weighted congestion games via best-responses. *Mathematics of Operations Research*, 47(1):643–664, 2022.

[96] V. Gkatzelis, K. Kollias, and T. Roughgarden. Optimal cost-sharing in general resource selection games. *Operations Research*, 64(6):1230–1238, 2016.

[97] R. Gopalakrishnan, J. R. Marden, and A. Wierman. Potential games are necessary to ensure pure Nash equilibria in cost sharing games. *Mathematics of Operations Research*, 39(4):1252–1296, 2014.

[98] R. L. Graham. Bounds for certain multiprocessing anomalies. *The Bell System Technical Journal*, 45(9):1563–1581, 1966.

[99] C. Hansknecht, M. Klimm, and A. Skopalik. Approximate pure nash equilibria in weighted congestion games. In *Proceedings of 17th International Workshop on Approximation Algorithms for Combinatorial Optimization Problems (APPROX)*, pages 242–257, 2014.

[100] T. Harks. Stackelberg strategies and collusion in network games with splittable flow. *Theory of Computing Systems*, 48(4):781–802, 2011.

[101] T. Harks and M. Klimm. On the existence of pure Nash equilibria in weighted congestion games. *Mathematics of Operations Research*, 37(3):419–436, 2012.

[102] T. Harks, S. Heinz, and M. E. Pfetsch. Competitive online multicommodity routing. *Theory of Computing Systems*, 45(3):533–554, 2009.

[103] B. Heydenreich, R. Müller, and M. Uetz. Mechanism design for decentralized online machine scheduling. *Operations Research*, 58(2):445–457, 2010.

[104] D. S. Hochbaum and D. B. Shmoys. A polynomial approximation scheme for scheduling on uniform processors: using the dual approximation approach. *SIAM Journal on Computing*, 17(3):539–551, 1988.

[105] R. Hoeksma and M. Uetz. The price of anarchy for minsum related machine scheduling. In *Proceedings of the 9th International Workshop on Approximation and Online Algorithms (WAOA)*, pages 261–273, 2011.

[106] N. Immorlica, L. E. Li, V. S. Mirrokni, and A. S. Schulz. Coordination mechanisms for selfish scheduling. *Theoretical Computer Science*, 410(17):1589–1598, 2009.

[107] T. Jelinek, M. Klaas, and G. Schäfer. Computing optimal tolls with arc restrictions and heterogeneous players. In *Proceedings of the 31st International Symposium on Theoretical Aspects of Computer Science (STACS)*, pages 433–444, 2014.

[108] G. Karakostas and S. G. Kolliopoulos. Stackelberg strategies for selfish routing in general multicommodity networks. *Algorithmica*, 53(1):132–153, 2009.

[109] P. Kleer and G. Schäfer. Tight inefficiency bounds for perception-parameterized affine congestion games. *Theoretical Computuer Science*, 754: 65–87, 2019.

[110] M. Klimm, D. Schmand, and A. Tönnis. The online best reply algorithm for resource allocation problems. In *Proceedings of the 12th International Symposium on Algorithmic Game Theory (SAGT)*, pages 200–215, 2019.

[111] K. Kollias. Nonpreemptive coordination mechanisms for identical machines. *Theory of Computing Systems*, 53(3):424–440, 2013.

[112] K. Kollias and T. Roughgarden. Restoring pure equilibria to weighted congestion games. *ACM Transactions on Economics and Computation*, 3(4): 21:1–21:24, 2015.

[113] E. Koutsoupias and C. Papadimitriou. Worst-case equilibria. In *Proceedings of the 16th Annual Conference on Theoretical Aspects of Computer Science (STACS)*, pages 404–413, 1999.

[114] J. Kulkarni and V. S. Mirrokni. Robust price of anarchy bounds via lp and fenchel duality. In *Proceedings of the 26th Annual ACM-SIAM Symposium on Discrete Algorithms (SODA)*, pages 1030–1049, 2015.

[115] V. S. A. Kumar and M. V. Marathe. Improved results for stackelberg scheduling strategies. In *Proceedings of the 29th International Colloquium on Automata, Languages and Programming (ICALP)*, pages 776–787, 2002.

[116] J. K. Lenstra, D. B. Shmoys, and É. Tardos. Approximation algorithms for scheduling unrelated parallel machines. *Mathematical Programming*, 46(3): 259–271, 1990.

[117] J. Y.-T. Leung, editor. *Handbook of scheduling - algorithms, models, and performance analysis*. CRC Press, 2004.

[118] J. Li. An $o(\log(n)/\log\log(n))$ upper bound on the price of stability for undirected Shapley network design games. *Information Processing Letters*, 109 (15):876–878, 2009.

[119] S. Li. Scheduling to minimize total weighted completion time via time-indexed linear programming relaxations. *SIAM Journal on Computing*, 49 (4), 2020.

[120] T. Lücking, M. Mavronicolas, B. Monien, and M. Rode. A new model for selfish routing. *Theoretical Computer Science*, 406(3):187–2006, 2008.

[121] K. Makarychev and M. Sviridenko. Solving optimization problems with dis-economies of scale via decoupling. *Journal of ACM*, 65(6):42:1–42:27, 2018.

[122] T. Mansour and M. Schork. *Commutation relations, normal ordering and Stirling numbers*. CRC Press, 2015.

[123] J. R. Marden and A. Wierman. Distributed welfare games. *Operations Research*, 61(1):155–168, 2013.

[124] J. Matouek and B. Gärtner. *Understanding and Using Linear Programming (Universitext)*. Springer-Verlag, Berlin, Heidelberg, 2006.

[125] R. Meir and D. Parkes. Playing the wrong game: smoothness bounds for congestion games with behavioral biases. *SIGMETRICS Performance Evaluation Review*, 43(3):67–70, 2015.

[126] R. Meir and D. C. Parkes. When are marginal congestion tolls optimal? In *Proceedings of the 9th International Workshop on Agents in Traffic and Transportation (ATT) co-located with the 25th International Joint Conference On Artificial Intelligence (IJCAI)*, 2016.

[127] V. S. Mirrokni and A. Vetta. Convergence issues in competitive games. In *Proceedings of the 7th International Workshop on Approximation Algorithms for Combinatorial Optimization Problems (APPROX)*, pages 183–194, 2004.

[128] D. Monderer and L. S. Shapley. Potential games. *Games and Economic Behavior*, 14(1):124–143, 1996.

[129] D. Mosk-Aoyama and T. Roughgarden. Worst-case efficiency analysis of queueing disciplines. In *Proceedings of the 36th International Colloquium on Automata, Languages and Programming (ICALP)*, pages 546–557, 2009.

[130] H. Moulin. The price of anarchy of serial, average and incremental cost sharing. *Economic Theory*, 36(3):379–405, 2008.

[131] H. Moulin and J. P. Vial. Strategically zero-sum games: The class of games whose completely mixed equilibria cannot be improved upon. *International Journal of Game Theory*, 7(3):201–221, 1978.

[132] U. Nadav and T. Roughgarden. The limits of smoothness: A primal-dual framework for price of anarchy bounds. In *Proceedings of the 6th International Workshop on Internet and Network Economics (WINE)*, pages 319–326, 2010.

[133] J. F. Nash. Equilibrium points in *n*-person games. *Proceedings of the National Academy of Science*, 36(1):48–49, 1950.

[134] N. Nisan, T. Roughgarden, É. Tardos, and V. V. Vazirani, editors. *Algorithmic Game Theory*. Cambridge University Press, 2007.

[135] F. Ordóñez and N. E. Stier-Moses. Wardrop equilibria with risk-averse users. *Transportation Science*, 44(1):63–86, 2010.

[136] M. J. Osborne. *An introduction to game theory*. Oxford University Press, 2004.

[137] D. Paccagnan and M. Gairing. In congestion games, taxes achieve optimal approximation. In *Proceedings of the 22nd ACM Conference on Economics and Computation (EC)*, pages 743–744, 2021.

[138] D. Paccagnan and J. R. Marden. Utility design for distributed resource allocation - part II: applications to submodular, covering, and supermodular problems. *IEEE Transactions on Automatic Control*, 67(2):618–632, 2022.

[139] D. Paccagnan, R. Chandan, and J. R. Marden. Utility design for distributed resource allocation - part I: characterizing and optimizing the exact price of anarchy. *IEEE Transactions on Automatic Control*, 65(11):4616–4631, 2020.

[140] D. Paccagnan, R. Chandan, B. L. Ferguson, and R. J. Marden. Optimal taxes in atomic congestion games. *ACM Transactions on Economics and Computation*, 9(3):19:1–19:33, 2021.

[141] R. Paes Leme, V. Syrgkanis, and É. Tardos. The curse of simultaneity. In *Proceedings of the 3rd Innovations in Theoretical Computer Science Conference (ITCS)*, pages 60–67, 2012.

[142] P. N. Panagopoulou and P. G. Spirakis. Algorithms for pure Nash equilibria in weighted congestion games. *Journal of Experimental Algorithmics*, 11 (2.7), 2006.

[143] C. H. Papadimitriou. Algorithms, games, and the internet. In *Proceedings of the 33rd Annual ACM Symposium on Theory of Computing (STOC)*, pages 749–753, 2001.

[144] C. H. Papadimitriou and T. Roughgarden. Computing correlated equilibria in multi-player games. *Journal of the ACM*, 55(3), 2008.

[145] M. Penn, M. Polukarov, and M. Tennenholtz. Congestion games with load-dependent failures: Identical resources. *Games and Economic Behavior*, 67 (1):156–173, 2009.

[146] M. Penn, M. Polukarov, and M. Tennenholtz. Asynchronous congestion games. In *Graph Theory, Computational Intelligence and Thought, Essays Dedicated to Martin Charles Golumbic on the Occasion of His 60th Birthday*, volume 5420, pages 41–53, 2009.

[147] M. Penn, M. Polukarov, and M. Tennenholtz. Random order congestion games. *Mathematics of Operations Research*, 34(3):706–725, 2009.

[148] M. Penn, M. Polukarov, and M. Tennenholtz. Taxed congestion games with failures. *Annals of Mathematics and Artificial Intelligence*, 56(2):133–151, 2009.

[149] M. Penn, M. Polukarov, and M. Tennenholtz. Congestion games with failures. *Discrete Applied Mathematics*, 159(15):1508–1525, 2011.

[150] A. C. Pigou. *The economics of welfare*. London: Macmillan and Co., 1938.

[151] G. Piliouras, E. Nikolova, and J. S. Shamma. Risk sensitivity of price of anarchy under uncertainty. *ACM Transactions on Economics and Computation*, 5(1):5:1–5:27, 2016.

[152] V. Ravindran Vijayalakshmi and A. Skopalik. Improving approximate pure nash equilibria in congestion games. In *Proceedings of the 16th International Conference of Web and Internet Economics (WINE)*, pages 280–294, 2020.

[153] V. Ravindran Vijayalakshmi, M. Schröder, and T. Tamir. Scheduling games with machine-dependent priority lists. *Theoretical Computer Science*, 855: 90–103, 2021.

[154] R. W. Rosenthal. A class of games possessing pure-strategy Nash equilibria. *International Journal of Game Theory*, 2(1):65–67, 1973.

[155] T. Roughgarden. The price of anarchy is independent of the network topology. *Journal of Computer and System Sciences*, 67(2):341–364, 2003.

[156] T. Roughgarden. Stackelberg scheduling strategies. *SIAM Journal on Computing*, 33(2):332–350, 2004.

[157] T. Roughgarden. The price of anarchy in games of incomplete information. In *Proceedings of the 13th ACM Conference on Electronic Commerce (EC)*, pages 862–879, 2012.

[158] T. Roughgarden. Intrinsic robustness of the price of anarchy. *Journal of ACM*, 62(5):32:1–32:42, 2015.

[159] T. Roughgarden and É. Tardos. How bad is selfish routing? *Journal of ACM*, 49(2):236–259, 2002.

[160] T. Roughgarden and É. Tardos. Bounding the inefficiency of equilibria in nonatomic congestion games. *Games and Economic Behavior*, 47(2):389–403, 2004.

[161] T. Roughgarden and É. Tardos. Bounding the inefficiency of equilibria in nonatomic congestion games. *Games and Economic Behavior*, 47(2):389–403, 2004.

[162] Tim Roughgarden and Florian Schoppmann. Local smoothness and the price of anarchy in splittable congestion games. *Journal of Economic Theory*, 156: 317–342, 2015.

[163] W. H. Sandholm. Pigouvian pricing and stochastic evolutionary implementation. *Journal of Economic Theory*, 132(1):367–382, 2007.

[164] A. S. Schulz and N. E. Stier-Moses. On the performance of user equilibria in traffic networks. In *Proceedings of the 14th Annual ACM-SIAM Symposium on Discrete Algorithms (SODA)*, pages 86–87, 2003.

[165] J. Sethuraman and M.S. Squillante. Optimal scheduling of multiclass parallel machines. In *Proceedings of the 10th Annual ACM-SIAM Symposium on Discrete Algorithms (SODA)*, pages 963–964, 1999.

[166] L. S. Shapley. *Additive and non-additive set functions*. PhD Thesis. Princeton University, 1953.

[167] D. B. Shmoys and É. Tardos. An approximation algorithm for the generalized assignment problem. *Mathematical Programming*, 62(1):461–474, 1993.

[168] C. Singh. Marginal cost pricing for atomic network congestion games. Technical report, 2008.

[169] M. Skutella. Convex quadratic and semidefinite programming relaxations in scheduling. *Journal of ACM*, 48(2):206–242, 2001.

[170] S. Suri, C. Tóth, and Y. Zhou. Selfish load balancing and atomic congestion games. *Algorithmica*, 47(1):79–96, 2007.

[171] C. Swamy. The effectiveness of stackelberg strategies and tolls for network congestion games. *ACM Transactions on Algorithms*, 8(4):36:1–36:19, 2012.

[172] N. K. Thang. Game efficiency through linear programming duality. In *Proceedings of the 10th Innovations in Theoretical Computer Science Conference (ITCS)*, pages 66:1–66:20, 2019.

[173] V. V. Vazirani. *Approximation Algorithms*. Springer, Berlin, Heidelberg, 2010.

[174] C. Vinci. Non-atomic one-round walks in polynomial congestion games. In *Proceedings of the 17th Italian Conference on Theoretical Computer Science (ICTCS)*, pages 11–22, 2016.

[175] C. Vinci. Non-atomic one-round walks in congestion games. *Theoretical Computer Science*, 764:61–79, 2019.

[176] P. von Falkenhausen and T. Harks. Optimal cost sharing for resource selection games. *Mathematics of Operations Research*, 38(1):184–208, 2013.

[177] J. von Neumann and O. Morgenstern. *Theory of games and economic behavior*. Princeton University Press, 1947.

[178] J. G. Wardrop. Some theoretical aspects of road traffic research. *Proceedings of the Institution of Civil Engineers, Part II*, 1(36):352–362, 1952.

Printed in the United States
by Baker & Taylor Publisher Services

Printed in the United States
by Baker & Taylor Publisher Services